John Jacobs Thomas

Farm Implements and Farm Machinery and the Principles of their Construction and Use

John Jacobs Thomas

Farm Implements and Farm Machinery and the Principles of their Construction and Use

ISBN/EAN: 9783337779542

Printed in Europe, USA, Canada, Australia, Japan

Cover: Foto ©berggeist007 / pixelio.de

More available books at **www.hansebooks.com**

FARM IMPLEMENTS

AND

FARM MACHINERY,

AND THE

Principles of their Construction and Use:

WITH

SIMPLE AND PRACTICAL EXPLANATIONS

OF THE

LAWS OF MOTION AND FORCE

AS APPLIED

ON THE FARM.

With 287 Illustrations.

BY

JOHN J. THOMAS.

NEW YORK:
ORANGE JUDD AND COMPANY,
245 BROADWAY.

PREFACE.

A small treatise,—the basis of the present work,—was originally published in the Transactions of the New York State Agricultural Society for 1850, under the title of "Agricultural Dynamics," or the Science of Farm Forces. A revised and greatly enlarged edition, adapted to general use, was afterwards issued in book form, with the name of "Farm Implements." Since the appearance of the earlier editions, great and rapid improvements have been made in farm machinery of nearly every kind; and the aim of the work in its present form is to supply, so far as its limits will admit, the information eagerly sought by cultivators in relation to all that has proved of value.

Another principal object has been to present in a simple and intelligible manner, the leading principles of Mechanical Science, applied directly in the farmer's daily routine, —that he may know the *reasons* of success and failure, and not be guided by random guessing. The first portion of the book is chiefly devoted to a practical explanation of these principles.

UNION SPRINGS, N. Y., 1869.

CONTENTS.

PART I.—MECHANICS.

CHAPTER I.
INTRODUCTION.—Value of Farm Machinery—Importance of a Knowledge of Mechanical Principles.................... 7–10

CHAPTER II.
GENERAL PRINCIPLES OF MECHANICS.—Inertia, Experiments and Examples—Inertia of Moving Bodies, or Momentum—Fast Riding—The Tiger's Leap—Pile Engines—Fly-wheel—Estimating the Quantity of Momentum—Compound Motion—Various Examples—Centrifugal Force.................... 10–22

CHAPTER III.
ATTRACTION.—Gravitation—Velocity of Falling Bodies—Resistance of the Air—Coin and Feather—Galileo's Famous Experiment—Cohesion—Soils—Strength of Materials—Capillary Attraction—The Earth a Desert without it—The Ascent of Sap—Centre of Gravity—Experiments—Upsetting Loads—Shouldering Bags—Rocking Bodies.................... 22–42

CHAPTER IV.
SIMPLE MACHINES, OR MECHANICAL POWERS.—Law of Virtual Velocities—The Lever—Many Examples of Levers—Estimating the Power of Levers—Three-horse Whiffle-tree—Compound Levers—Weighing Machines—Stump Pullers—A Wild Theory—Wheel and Axle—Examples—Band and Cog-work—The Pulley—Packer's Stone Lifter—The Inclined Plane—Crooked Roads—Power of Locomotives—Good and Bad Roads—The Wedge—The Screw—Knee-joint Power—Lever Washing Machine—Cheese Presses—Rolling Mills—Straw Cutters........ 42–74

CHAPTER V.
APPLICATION OF MECHANICAL PRINCIPLES in the Structure of Implements and Machines—Various Examples—Calculating the Strength of Parts.................... 75–81

CHAPTER VI.
FRICTION.—How to ascertain its Amount—Friction on Roads—Resistance of Mud—The Results of the Dynamometer—Width of Wheels—Velocity—Size of Wheels on Roads—Friction Wheels—Lubricating Substances—Friction necessary to Existence.................... 81–93

CHAPTER VII.
PRINCIPLES OF DRAUGHT—Applied to Wagons—To Plows—Combined Draught of Animals—Whiffle-trees for Three Horses—

Potter's do.—Wier's Single-tree—The Dynamometer—Self-registering do.—Waterman's do.—Dynamometer for Rotary Motion.. 93–108

CHAPTER VIII.

APPLICATION OF LABOR.—Power of Horses—Of Men—Best Way to Apply Strength..108–113

CHAPTER IX.

MODELS OF MACHINES.—Common Blunders—Works of Creation Free from Mistakes....................................113–115

CHAPTER X.

CONSTRUCTION AND USE OF FARM IMPLEMENTS AND MACHINES —IMPLEMENTS OF TILLAGE.—Importance of Simplicity— PLOWS—Rude Specimens—Cast-iron and Steel do.—Character of a Good Plow—The Cutting Edge—Mould-board— Easy Running Plows—Crested Furrow Slices—Lapping and Flat Furrows—How to Plow Well—Fast and Slow Plowing —The Double Michigan—The Subsoil Plow—The Paring Plow—Gang Plow—Ditching Plow—Mole Plow—Coulters —Weed Hook and Chain—PULVERIZERS—Harrows—Geddes' Harrows—Scotch do.—Morgan Harrow—Norwegian do.—Shares' do.—Cultivators—Holbrook's—Alden's—Garrett's Horse-hoe—Two-horse Cultivators—Sulky do.—Comstock's Spader—Clod Crushers—Roller...................115–152

CHAPTER XI.

SOWING MACHINES—Wheat Drills—Bickford and Huffman's do.— Seymour's Broadcast Sower—Corn Planters—True's Potato Planter—Hand Drills.............................152–157

CHAPTER XII.

MACHINES FOR HAYING AND HARVESTING.—Mowing and Reaping Machines — Cutter-bar — Combined Machines — Self Rakers—Johnson's do.—Marsh's and Kirby's do.—Dropper —Binders—Marsh's Harvester—Durability and Selection of Machines—Hay Tedders—Bullard's do.—American do.— Horse Rakes—Revolving do.—Sulky Revolvers—Warner's do. — Spring Tooth Rakes — Hollingsworth's do. — Hay Sweep—Horse Forks—Gladding's do.—Palmer's, Myers' Beardsley's, Raymond's—Harpoon Forks—Hay Carriers— Hicks' do.—Building Stacks—Palmer's Hay Stacker—Raymond's Hay Stacker—Dederick's Hay Press—Beater do.— Hay Loaders..158–186

CHAPTER XIII.

THRASHING, GRINDING, AND PREPARING PRODUCTS. — Value of Thrashing Machines — Endless Chain Power — How to Measure Power of—Churning by Tread Power—Pitt's Elevator—Corn Shellers—Burralls', Richards'—Root Washer—Root Slicers—Farm Mills, Allen's, Forsman's—Emery Cotton Gin..186-197

PART II.—MACHINERY IN CONNECTION WITH WATER.
CHAPTER I.
HYDROSTATICS.—Upward Pressure—Measuring Pressure—Calculating Strength of Tubes, etc.—Artesian Wells—Determining Pressure in Vessels—A Puzzle Explained—Hydrostatic Bellows—Press—Specific Gravities—Table of do.—Weight and Bulk of a Ton of Different Substances........198-210

CHAPTER II.
HYDRAULICS.—Velocity of Water—Discharge of Water through Pipes—Velocity in Ditches—Leveling Ditches—Archimedean Screw-pumps—For Cisterns—Non-freezing do.—For Deep Wells—Drive Pumps—Chain Pumps—Rotary do.—Suction and Forcing Pump—Turbine Water Wheels—The Water Ram—Water Engines for Gardens—Flash Wheel—Nature of Waves—Size of do.—Preventing Inroads by do.—Cisterns—To Determine Contents of...................211-238

PART III.—MACHINERY IN CONNECTION WITH AIR.
CHAPTER I.
PRESSURE OF AIR.—Weight of the Atmosphere—Hand Fastened by Air—Barometer—Measuring Heights—Syphon.........239-245

CHAPTER II.
MOTION OF AIR.—Winds—Wind-mills, how Used—Brown's do.—Causes of Wind—Chimney Currents—Construction of Chimneys—To Cure Smoky do.—Chimney Caps—Ventilation...245-259

PART IV.—HEAT.
CHAPTER I.
Conducting Power—Expansion, Great Force of—Experiments with—Steam Engine—do., for Farms—Steam Plows—Latent Heat—Green and Dry Wood......................260-276

CHAPTER II.
RADIATION.—Several Examples in Domestic Economy—Dew and Frost—Frost in Valleys—Sites for Fruit Orchards........276-280

APPENDIX.
Apparatus for Experiments.................................281-283
Discharge of Water through Pipes........................... 284
Velocity of Water in Pipes................................. 284
Rule for Discharge of Water................................285-286
Velocity of Water in Tile Drains........................... 286
Glossary...287-296

FARM IMPLEMENTS

AND

FARM MACHINERY.

PART I.

MECHANICS.

CHAPTER I.

INTRODUCTION.

No farm can be well furnished without a large number of machines and implements. Scarcely any labor is performed without their assistance, from the simple operations of hoeing and spading, to the more complex work of turning the sod and driving the thrashing-machine. The more perfect this machinery, and the better fitted to its work, the greater will be the gain derived by the farmer from its use. It becomes, therefore, a matter of vital importance to be able to construct the best, or to select the best already constructed, and to apply the forces required for the use of such machines to the greatest possible advantage.

Nothing shows the advancement of modern agriculture in a more striking light than the rapid improvement in farm implements. It has enabled the farmer within the last fifty years to effect several times the work with an equal force of horses and men. Plows turn up the soil deeper, more evenly and perfectly, and with greater ease of draught; hoes and spades have become lighter and more efficient; grain, instead of being beaten out by the slow and laborious work of the flail, is now showered in torrents from the thrashing-machine; horse-rakes accomplish singly the work of many men using the old hand-rake; horse-forks convey hay to the barn or stack with ease and rapidity; twelve acres of ripe grain are neatly cut in one day with a two-horse reaper; grain drills and planting machines, avoiding the tiresome drudgery of hand labor, distribute the seed for the future crop with evenness and precision.

The owner of a seventy-thousand-acre farm in Illinois carries on nearly all his work by labor-saving machinery. He drives posts by horse-power; breaks his ground with Comstock's rotary spader; mows, rakes, loads, unloads, and stacks his hay by horse-power; cultivates his corn with two-horse, seated or sulky cultivators; ditches low ground, sows and plants by machinery; so that his laborers ride in the performance of their tasks without exhausting their strength with needless walking over extended fields.

The great value of improved farm machinery to the country at large has been lately proved by the introduction of the reaper. Careful estimate determined that the number of reaping machines introduced throughout the country up to the beginning of the great rebellion, performed an amount of labor while working in harvest nearly equal to a million of men with hand implements. The reaper thus filled the void caused by the demand on workingmen for the army. An earlier occurrence of that war must therefore have resulted in the general ruin of the grain interest, and prevented the annual shipment of the millions during that gigantic contest, which so greatly surprised the commercial savans of Europe.

The implements and machines which every farmer must have who does his work well are numerous and often costly. A farm of one hundred acres requires the aid of nearly all the following; two or more good plows, a shovel-plow, a small plow, a subsoiler, a single and two-horse cultivator, a seed-planter, a grain-drill, a roller, a harrow, a fanning-mill, a straw-cutter, a root-slicer, a farm wagon with hay-rack, an ox-cart, a horse-cart, wheel-barrow, sled, shovels, spades, hoes, hay-forks and manure-forks, hand-rakes and horse-rakes, scythes and grain-cradle, grain-shovel, maul and wedges, pick, axes, wood-saw, hay-knife, apple-ladders, and many other smaller conveniences. The capital for furnishing the farms in the Union has been computed to amount to more than five hundred millions of dollars, and as much more is estimated to be yearly paid for the labor of men and horses throughout the country at large. To increase the effective force of labor only one-fifth would, therefore, add annually one hundred millions in the aggregate to the profits of farming.

A knowledge of the science of mechanics is not so well understood among all classes of people as it should be. A loss often occurs from the want of a correct knowledge of mechanical principles. The strength of laborers is badly applied by the use of unsuitable tools, and that of teams is partly lost by being ill adjusted to the best line of draught. We may perhaps see but few instances of so great a blunder as the ignorant teamster committed who fastened his smaller horse to the shorter end of the whiffle-tree, to balance the large horse at the longer end; yet instances are not uncommon where operations are performed to almost as great a disadvantage, and which, to

1*

a person well versed in the science of mechanics, would appear nearly as absurd.

It is well worth while to look at the achievements made through a knowledge of mechanical principles. Compare the condition of barbarous and savage tribes with that of modern civilized nations. The former, scattered in comfortless hovels, subsist by precarious hunting, or on scanty crops raised on patches of ground by means of the rudest tools. The latter are blessed with smooth, cultivated fields, green meadows, and golden harvests. Commerce with its hum of business, extending through populous cities, and along a hundred far-stretching lines of rail-ways, scatters comforts and luxuries to millions of homes; while ships for foreign commerce thread every channel and whiten every sea. The contrast exhibits the difference between ignorance on the one hand, and the successful application of scientific principles on the other. It is our present object to point out to the farmer the advantages which would result from a wide extension, through all classes, of this knowledge, that the opportunities may be continually increased for general improvement.

CHAPTER II.

GENERAL PRINCIPLES OF MECHANICS.

Having briefly pointed out some of the advantages to the farmer of understanding the principles of the machines he constantly uses, we now proceed to an examination of these principles. It will be most convenient to begin with the simpler truths of the science, proceeding, as we advance, to their application in the construction of **machines.**

INERTIA.

An important quality of all material bodies is INERTIA. This term expresses their passive state—that is, that no body (not having life), when at rest, can move itself, nor, when in motion, can stop itself. A stone has not power to commence rolling of its own accord; a carriage can not travel on the road without being drawn; a train of cars never commences gliding upon the rails without the power of the locomotive.

On the contrary, a body, when once set in motion, will continue in motion perpetually, unless stopped by something else. A cannon ball rolled upon the ground moves on until its force is gradually overcome by the resistance of the rough earth. If a polished metallic globe were driven swiftly on a level and polished metallic plane, it would continue in motion a long time and travel to a great distance; but still the extremely minute roughness of the surfaces, with the resistance of the air, would continually diminish its speed until finally stopped. A wheel made to spin on its axis revolves until the friction at the axis and the impeding force of the air bring it to rest. But if the air is first removed,

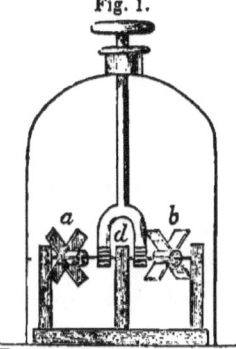

Fig. 1.

Fans revolving in a vacuum.

by means of an air-pump, the motion will continue much longer. Under a glass receiver, thus exhausted, a top has been made to spin for hours, and a pendulum to vibrate for a day. The resistance of the air may be easily perceived by first striking the edge and then the broad side of a large piece of pasteboard against the air of a room. It is further shown by means of an interesting experiment with the air-pump. Two fan-wheels, made of sheet tin, one, *a*, striking the air with its edges, and the other, *b*, with its broad faces (fig.

1), are set in motion alike; *b* is soon brought to rest, while *a* continues revolving a long time. If now they are placed under the receiver of an air-pump, the air exhausted, and motion given to them alike by the rack-work *d*, they will both continue in motion during the same period.

There is no machinery made by man free from the checking influence of friction and the air; and for this reason, no artificial means have ever devised a perpetual motion by mechanical force. But we are not without a proof that motion will continue without ceasing when nothing operates against it. The revolutions of the planets in their orbits furnish a sublime instance; where removed from all obstructions, these vast globes wheel around in their immense orbits, through successive centuries, and with unerring regularity, preserving undiminished the mighty force given them when first launched into the regions of space.

To set any body in motion, a *force* is requisite, and the heavier the body, the greater must be the force. A small stone is more easily thrown by the hand than a cannon ball; speed is more readily given to a skiff than to a large and heavy vessel; but the same force which sets a body in motion is required to stop it. Thus a wheel or a grindstone, made to revolve rapidly, would need as great an effort of the arm to stop it suddenly as to give it sudden motion. An unusual exertion of the team is necessary in starting a loaded wagon; but when once on its way, it would require the same effort of the horses to stop it as to *back* it when at rest.

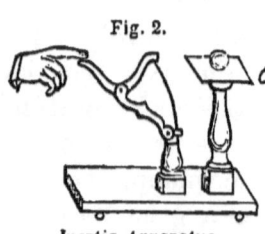

Fig. 2.

Inertia Apparatus.

The force of inertia is finely exhibited by means of a little instrument called the *inertia apparatus* (fig. 2). A marble or small ball is placed on a card, *c*, resting on a concave stand. A spring snap is then made to strike the

INERTIA.—EXPERIMENTS AND EXAMPLES.

card, throwing it to a distance, but leaving the ball upon the hollow end of the stand. The same experiment may be easily performed by placing a very small apple or other solid on a card, the whole resting on a common sand-box, or even the hollow of the hand. A sudden snap with the finger will throw the card away, while the apple will drop into the cavity. The following experiment is still more striking: Procure a thread just strong enough to bear three pounds, and hang upon it a weight of two pounds and a half. Another half pound would break it. Now tie another thread, strong enough to bear one pound, to the lower hook of the weight. If the lower thread be pulled *gradually*, the *upper* thread will of course break; but if it be pulled with a *jerk*, the *lower* thread will break.

Fig. 3.

If the jerk be very sudden, the lower string will break, even it be considerably stronger than the upper, the *inertia* of the weight requiring a great force to overcome it suddenly. The threads used in this experiment may be easily had of any desired strength by taking the finest sewing cotton, and doubling to any desired extent.

This experiment shows the reason why a horse, when he suddenly starts with a loaded wagon, is in danger of breaking the harness; and why a heavier weight may be lifted with a windlass or pulley having a weak rope, if the strain is gradual and not sudden. For the same reason, glass vessels full of water are sometimes broken when hastily lifted by the handle. When a bullet is fired through a pane of glass, the inertia retains the surrounding glass in its place during the moment the ball is passing, and a round hole only is made; while a body moving more slowly, and pressing the glass for a longer space of time, fractures the whole pane.

INERTIA OF MOVING BODIES, OR MOMENTUM.

Momentum is the *inertia of a moving body.* When a force is applied to a heavy body, its motion is at first slow; but the little momentum it thus acquires, added to the applied force, increases the velocity. This increase of velocity is of course attended with increased momentum, which again, added to the acting force, still further quickens the speed. For this reason, when a steam-boat leaves the pier, and its paddle-wheels commence tearing through the water, the motion, at first slow, is constantly accelerated until the increasing resistance of the water becomes equal to the strength of the engine and the momentum.* Were it not for the momentum of moving bodies (inertia existing), no speed ever could be given to any heavy body, as a carriage, boat, or train of cars.

The chief danger in fast riding, or fast traveling of any kind, is from the momentum given to the traveler. If a rail-way passenger should step from a car when in full motion, he would strike the earth with the same velocity as that of the train; or if the train at thirty miles an hour should be instantly stopped, the passengers would be pitched forward with a swiftness equal to thirty miles an hour. When a horse suddenly stops, the momentum of the rider tends to throw him over the horse's head. When a wagon strikes an obstruction, the driver falls forward. A case in court was once decided against the plaintiff, who claimed that the defendant had driven against his wagon with such force as to throw the plaintiff to a great distance; but the fact was shown that by such momentum he himself must have been driving furiously, and not the defendant, and he lost his suit.

---

* In ordinary practice, this is not strictly correct, as *friction* will make some difference. This influence will be more particularly considered on a subsequent page. Its omission here does not at all alter the *principle* under consideration.

An Eastern traveler once succeeded in saving his life by a ready knowledge of this principle. He was closely pursued by a tiger, and when near a precipice, watching his opportunity, he threw his coat and hat on a bush, and jumped one side, when the animal, leaping swiftly on the concealed bush, was carried by momentum over the precipice.

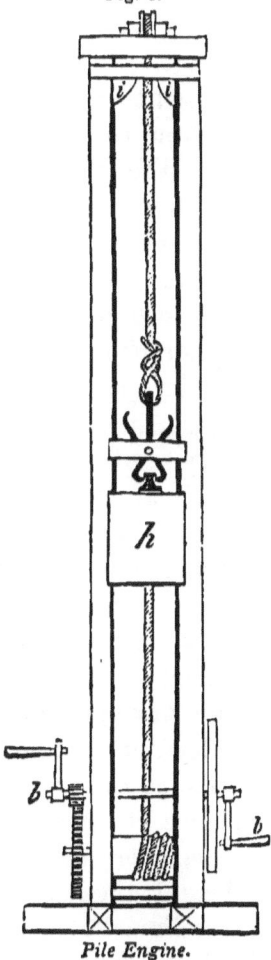

Fig. 4.

Pile Engine.

As a large or heavy body possesses greater momentum than a small or light one, so any body moving with great speed possesses more than one moving slowly; for instance, the momentum of a rifle ball is so great as to carry it through a thick plank, while, if thrown slowly, it would scarcely indent it.

This property of bodies is applied with great advantage to many practical purposes. The momentum of the hammer drives the nail into the wood; for the mere pressure of its weight would not do it, if it were a hundred times as heavy. Wedges are driven by employing the same kind of power.

On a larger scale, the pile-engine operates in a similar manner. The ram or weight, h (fig. 4), is slowly lifted by means of a pulley and wheel-work, worked by the handles or cranks, $b\ b$, until the arms of the tongs which hold the ram are compressed in the cheeks, $i\ i$, when it suddenly falls with prodigious force on the pile or post to be driven. In this way long posts of great size are forced into the mud of swamps and river bottoms, where other means would fail. When a steam-engine is used for lifting the ram, the work is more rapidly performed.

An interesting example of the use and efficiency of momentum is furnished by the *water-ram*, a machine for raising water, described on a subsequent page.

The *fly-wheel*, a large and heavy wheel used to regulate the motion of machinery, derives its value from the power of inertia, or momentum, which prevents the machine from stopping suddenly when it meets with any unusual obstruction. In the common thrashing-machine, it has been found that a heavy cylinder, by acting as a fly-wheel, renders the motion steadier, and less liable to become impeded by large sheaves of grain. An ignorance of this principle has sometimes proved a serious inconvenience. A farmer, having occasion to raise a large quantity of water, erected a horse-pump; but at every stroke of the pump the animal was suddenly thrown loosely forward, and again jerked backward, as the piston fell lightly and rose heavily. A fly-wheel attached to the machinery would have prevented this unpleasant jerking, and have enabled the horse, in consequence, to accomplish more work. In the pile-driving engine, where a great weight is suddenly thrown loose from a height, the horses would be pitched forward when suddenly relieved of this load but for the regulation of a fly-wheel, the motion of which is not quickly changed, neither from fast to slow nor from slow to fast.

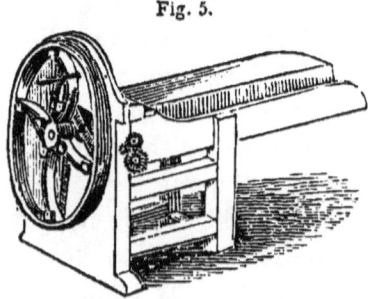

Fig. 5.

Straw-cutter with fly-wheel.

Where there is a rapid succession of forces required in practice, the fly-wheel is usually of great advantage. Hence its use in all revolving straw-cutters, where the knives make quickly-repeated strokes (fig. 5). More recently it has been applied to the dasher-churn (fig. 6), where the rapid upright strokes are so well known to be very fatiguing for the amount of force applied.

By thus regulating motion, the fly-wheel frequently enables an irregular force to accomplish work which otherwise it could not perform. Thus a man may exert a force equal to raising a hundred pounds, yet, when he turns a crank, there is one part of the revolution where he works to great disadvantage, and where his utmost force will not balance forty pounds. Hence, if the work is heavy, he may not be able to turn the crank, nor to do any work at all. If, however, a fly-wheel be applied, by gathering force at the most favorable part of the turning, it carries the crank through the other part.

Fig. 6.

Churn with a fly-wheel, for equalizing the motion.

An error is sometimes committed by supposing the fly-wheel actually creates power, for as much force is required to give it momentum as it afterward imparts to the machine; it consequently only accumulates and regulates power.

On rough roads, the force of inertia causes a severe strain to a loaded wagon when it strikes a stone. The horses are chafed, the wagon and harness endangered, and the load jarred from its place. This inconvenience is avoided in part by placing the box upon springs, which, by yielding to the blow, gradually lessen the effects of the shock. For carts and slowly moving lumber-wagons springs are useful, but more so as the velocity and momentum increase. Even on so smooth a surface as a rail-road, it was found by experiments made some years ago, that when the machinery of a locomotive was placed upon springs, it would endure the wear and tear of use four times as long as without them.

For this reason, a ton of stone, brick, or of sand, is harder for a team than a ton of wool or hay, which possesses considerable elasticity.

ESTIMATING THE QUANTITY OF MOMENTUM.

The *quantity* of momentum is estimated by the velocity and weight of the body taken together. Thus a ball of two pounds weight moves with twice the force of a one-pound ball, the speed being equal; a ten-pound ball with ten times the force, and so on. A body moving at the rate of two feet per second possesses twice the momentum of another of equal size with a velocity of only one foot per second. A musket ball, weighing one ounce, flying with fifty times the speed of a cannon ball, weighing fifty ounces, would strike any object with equal force; or, if they should meet each other from opposite directions, the momentum of both would be mutually destroyed, and they would drop to the earth.

Where the mass is very great, even if the motion is slow, the momentum is enormous. A large ship floating near a pier wall may approach it with so small a velocity as to be scarcely perceptible, and yet the force would be enough to crush a small boat. When great weight and speed are combined, as in a rail-way locomotive, the force is almost irresistible. This circumstance often insures the safety of the passengers; for as nothing is capable of instantly overcoming so powerful a momentum, when accidents occur the speed is more gradually slackened, and the passengers are not pitched suddenly forward. A light wagon, rapidly driven, possessing but little comparative force, is more suddenly arrested, and the danger is greater.

When two bodies meet from opposite directions, each

sustains a shock equal to the united forces of both. Two men accidentally coming in contact, even if walking moderately, receive each a severe blow; that is, if each were walking three miles an hour, the shock would be the same as if one at rest were struck by the other with a velocity of six miles an hour. This principle accounts for the destructive effects of two ships running foul of each other at sea, or of the collision of two opposite trains on a rail-road.

The preceding principles show that a sledge, maul, or axe will always strike more effective blows when made heavier, if not rendered unwieldy.

COMPOUND MOTION.

It often happens that two or more forces act on the same body at the same time. If they all act in the same direction, the effect will be equal to the sum of the forces taken together; but if they act in opposite directions, the forces will tend to destroy each other. If two equal forces act in contrary directions, they will be completely neutralized, and no motion will be produced. Thus, as an example of these forces—a bird flying at the rate of forty miles an hour, *with* a wind blowing forty miles an hour, will be driven onward by these two combined forces eighty miles an hour; but if it undertake to fly against such a wind, it will not advance at all, but remain stationary. A similar result will take place if a steam-boat, having a speed of ten miles an hour, should first run down a river with a current of equal velocity, and then upward against the current; in the first case it would move twenty miles an hour, and in the latter it would not move at all.

Where forces act neither in the same nor in opposite directions, but obliquely, the result is found in the following manner: If a ball, placed at the point a (fig. 7), be struck by two different forces at the same moment, in the direction shown by the two arrows, and if one force be just sufficient to carry it from a to c, and the other to carry it from a to b, then it will move intermediate between the two, in the direction of the diagonal of the parallelogram $a\ d$, and to a distance just equal to the length of this diagonal or cross-diameter.

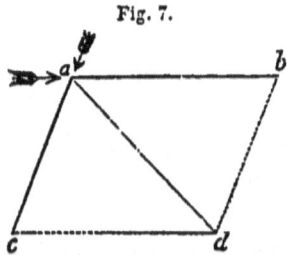

Fig. 7.

When the forces act very nearly together, the parallelogram of the forces will be very narrow and quite long, with a long diagonal (fig. 8); but if they act on nearly opposite sides of the ball, they will very nearly neutralize each other, and the diagonal or result will be very short, showing that the motion given to the ball will be very small (fig. 9).

Fig. 8.

These examples show the importance of having teams attached to a plow or to a wagon very nearly in a straight line with the draught, or else a part of the force will be lost; and also the importance, when several animals are drawing together, of their working as nearly as possible in the same straight line. For, the more such forces deviate from a right line, the more they will tend to destroy or neutralize each other.

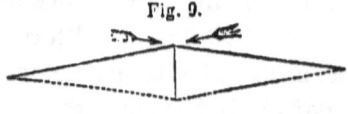

Fig. 9.

A familiar example of the result of two oblique forces is furnished when a boat is rowed across a river. If the river has no current, the boat will pass directly from bank to bank perpendicularly; but if there is a current, its track will form a diagonal, and it will strike the opposite bank lower down, according to the rapidity of the stream and the slowness of the boat.

Another instance is afforded when a ferry-boat is anchored, by means of a long rope, to a point some distance above (fig. 10); the boat, being turned obliquely, will pass from one bank to the other by the force of the current. Here the water tends to carry the boat downward, while the force of the rope acts upward; the boat passes between the two from bank to bank.

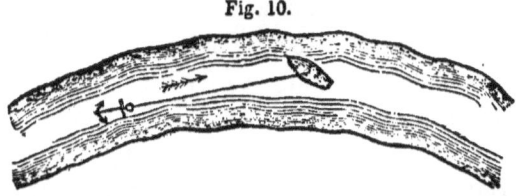
Fig. 10.

The ascent of a kite is precisely similar, the wind and the string being counteracting forces. When a vessel sails under a side wind, the resistance of the keel against the water, and the force of the wind against the sail, act in different directions, and produce a motion of the vessel between them.

CENTRIFUGAL FORCE.

All bodies, when in motion, have a tendency to move forward in a straight line. A stone thrown into the air is gradually bent from this straight course into a curve by the attraction of the earth. When a ball is shot from a gun, the force being greater, it flies in a longer and straighter curve. A familiar example also occurs, while driving a wagon rapidly, in attempting to turn suddenly to the right or left; the tendency of the load to move straight on will sometimes cause its overthrow. An observance of this principle would prevent the error which some commit by making sharp turns or angles in ditches and water-courses; the onward tendency of the water drives it against the bank, checks its course, and wears away the earth. By giving the ditch a curve, the water is but slightly impeded, and a much larger quantity will escape through a channel of any given size.

When a grindstone is turned rapidly, the water upon its surface is thrown off by this tendency to move in straight lines. In the same way, a weight fastened to a cord, whirled by the hand, will keep the cord stretched during the revolution. A cup of water, attached to a cord, may be swung over the head without spilling, the water being held by centrifugal force. The same principle causes a stone, when it leaves a sling, to fly off in a line. This tendency to fly off from a revolving centre is called *centrifugal force*—the word *centrifugal* meaning *flying from the centre*. Large grindstones, driven with great velocity by machinery, are sometimes split asunder by centrifugal force.

The most sublime examples of this force occur in the motion of the earth and planets, which will be more fully explained in a future page.

CHAPTER III.

ATTRACTION.

GRAVITATION.

The earth, as is well known, is a mass of matter in the form of a globe, the diameter being upward of 7900 miles. From its enormous size and the small portion seen from one point, the surface appears flat, except where broken into mountains and valleys.

The tendency which all bodies possess of falling toward the earth is owing to the attractive force which this great mass of matter exerts upon them. This kind of attrac-

GRAVITATION.—VELOCITY OF FALLING BODIES. 23

tion is called *gravitation*. The force with which a body is thus drawn is the *weight* of that body.

When a stone is dropped from the hand, its velocity is at first slow, but continues to increase till it strikes the earth; hence, the further it falls the harder it will strike. This accelerated motion is precisely similar to that of a steam-boat when it first leaves the wharf; the force of gravity may be compared to the driving power of the engine, and the quickened velocity of the falling stone to the increased headway of the boat.

All bodies, whether large or small, fall equally fast, unless they are so light as to be borne up in part by the resistance of the air. In the first second of time they fall 16 feet; in the second, 3 times 16, or 48 feet; in the third second, 5 times 16, or 80 feet, and so on. Or, if the whole distance fallen be taken together, they fall 16 feet in one second, 4 times 16 in two seconds, 9 times 16 in three seconds, and so forth. In other words, the whole distance is equal to the square of the time. This is plainly exhibited in the following table:

Seconds, from beginning to fall.	1	2	3	4	5	6
Whole height fallen in feet.	16	4×16 or 64.	9×16 or 144.	16×16 or 256.	25×16 or 400.	36×16 or 576.
Space fallen in each second in feet.	16	3×16 or 48.	5×16 or 80.	7×16 or 112.	9×16 or 144.	11×16 or 176.

A stone or other body will fall 1 foot in a fourth of a second, 3 feet the next fourth, 5 feet the third fourth, and 7 feet the last fourth; which is the same as 4 feet in half a second, 9 feet in three-fourths of a second, and 16 feet for the whole second.

The depth of an empty well, or the height of a precipice, may be nearly ascertained by observing the time required for the fall of a stone to the bottom. The time may be measured by a stop-watch, or, in its absence, a pendulum may be made by fastening a pebble to a cord, which will swing from the hand in regular vibrations of an exact second each if the cord be 39⅛ inches long, or of half a second each if it be about 9¾ inches long.

The velocity increases simply as the time, that is, the speed in falling is twice as great in two seconds as in one; three times as great in three seconds; four times as great in four seconds, and so forth. A stone will fall four times as far in two as in one second, while its velocity will be doubled; nine times as far in three seconds, while its velocity will be tripled, etc.

If a stone is thrown upward, its motion continues gradually to decrease, at the same rate as it increases in falling; hence the same time is required to reach its highest point, as to fall from that point back to the earth. Therefore the velocity with which it is first projected upward is equal to the velocity which it attains at the moment of striking the ground. There is an exception, however, to this general rule. In a vacuum it would be perfectly correct, but in ordinary practice the resistance of the air tends to diminish the velocity while ascending, and still further to retard it while descending. For this reason, it will fall with less speed than it first arose. For heavy bodies and small distances, this exception would be imperceptible; but with small bodies falling from great heights, the difference will be considerable.

The velocity of a stone after falling one second, or sixteen feet, is at the rate of thirty-two feet per second; hence, if thrown upward at that rate, it will rise just sixteen feet high. After falling three seconds, the rate is ninety-six feet; and hence, if projected upward at ninety-six feet per second, it will rise nine times sixteen feet, or one hundred and forty-four feet high. And so of other heights.

Were it not for the resistance of the air, a feather would fall as swiftly as a leaden ball. This is conclusively shown by an interesting experiment. A tall glass jar (fig. 11), open at the bottom, is covered with a brass cap, fitting it air-tight. Through this cap passes an air-tight wire, which, by turning, opens a small pair of pincers. Within these are placed a feather and a half dollar, and the air is then thoroughly drawn from the receiver by means of an air-pump. The wire is turned, and the feather and coin both drop at once, and strike the bottom at the same moment.

There are many examples showing the accelerated motion and increased force of falling bodies. When a large stone rolls down a mountain, it first moves slowly, but afterwards bounds with rapidity, sweeping before it all smaller obstacles. Hailstones, although small, acquire such velocity as to break windows; and but for the resistance of the air, they would be much more destructive. The blow of a sledge-hammer is more severe as it is lifted to a greater height. Newton was first led to the examination of the laws of gravity by observing, when sitting under an apple-tree, that the fruit struck his hand with greatest severity when it fell from the top of the tree.

Fig. 11.

Feather and coin falling alike in a vacuum.

It is not an unusual error to suppose that a large body will fall more rapidly than a small one. Some can scarcely believe that a fifty-six pound weight will not drop with a greater velocity than a small nail, not remembering that a proportionately greater force is required to overcome the inertia and set the larger body in motion. This error existed for many centuries, from the time of Aristotle until Galileo first questioned its correctness. The celebrated experiment which established the truth on this subject, and led to the discovery of the laws of falling bodies just explained, and which formed an era in modern philosophy, was performed from the top of the leaning tower of Pisa. Galileo was a philosophical teacher, and, being a man who thought for himself, soon discovered, by reasoning, the errors that had been received without a doubt for more than twenty centuries. All the learning of the age and the wisdom of the universities were against him, and in favor of this time-honored error, the truth of which no one had ever thought of submitting to experiment. The hour of trial arrived, when he, an obscure young man, stood nearly alone on one side, while the multitude, with all the power and confessed knowledge of the age, were on the other.

The balls to be employed were carefully weighed and scrutinized to detect deception, and the parties were satisfied. The one ball was exactly twice the weight of the other. The followers of Aristotle maintained that when the balls were dropped from the top of the tower, the heavy one would reach the ground in exactly half the time employed by the lighter ball. Galileo asserted that the weights of the balls would not affect their velocities, and that the times of descent would be equal. The balls were conveyed to the summit of the lofty tower—the crowd assembled round the base—the signal was given—the balls were dropped at the same instant, and swiftly descending, at the same moment struck the earth. Again and again the experiment was repeated with uniform results. Galileo's triumph was complete—not a shadow of doubt remained; but, instead of receiving the congratulations of honest conviction, private interest, the loss of place, and the mortification of confessing false teaching, proved too strong for the candor of his adversaries. They clung to their former opinions with the tenacity of despair, and he was driven from Pisa.*

---

* Mitchell's Lectures.

COHESION.

The attraction of gravitation, as we have just seen, takes place between bodies at a greater or less distance from each other. There is another kind of attraction, acting only when the parts of substances are in *actual contact;* this is called *cohesion.* It is this which holds the parts of a body together and prevents it from falling to pieces. It may be shown by taking two pieces of lead, and, after having made upon them two smoothly-shaven surfaces with a knife, pressing them firmly together with a twisting motion (fig. 14). The asperities of the surfaces are thus pushed down, and the particles are brought into close contact, so that cohesion immediately takes place between them, and some force will be required to draw them asunder.* Two pieces of melted wax adhere together in the same way. Melted pitch or other similar substance, smeared thinly over the polished surfaces of metal or glass, also causes cohesion to take place between them. Smooth iron plates, two inches in diameter, have been made to stick together so firmly with hot grease as to require, when cold, a weight of half a ton to draw them apart. Plates of brass of the same size, cemented by means of pitch, required 1400 pounds. On this principle depends the efficacy of those substances which are used for cementing broken vessels.

Fig. 14.

Cohesive attraction in two lead balls.

The most perfect artificial polish which can be given to hard metals is still so rough as to prevent the faces from

---

*.That this is not atmospheric pressure, like that which holds two panes of wet glass together, is shown by the fact that it requires nearly as great a force to separate them when they are placed under the exhausted receiver of an air-pump. Besides this, atmospheric pressure is much weaker than this force, with so small a surface.

coming into close contact; hence they must be either melted, or softened like iron when it is welded.

The different degrees of cohesion which take place between the particles of various soils, to reunite them after they have been crumbled asunder, occasion the main difference between light and heavy soils. When a light soil becomes soaked with water, the particles adhere in a very slight degree; and hence, when it becomes dry again, it is easily worked mellow. But if it be of a clayey nature, too much moisture softens it like melted wax: the particles are thus brought into close contact, and strong adhesion takes place; hence the hardness and difficulty of working such soils when again dried. This adhesion is lessened by applying sand, chip-dirt, straw, yard-manure, or by burning the earth, but more especially by thorough draining, which, preventing the clay from becoming so moist and soft, lessens the adhesion of its parts.

Different substances are hard, soft, brittle, or elastic, according to the different degrees or modes of action in the attraction of cohesion.

STRENGTH OF MATERIALS.

It is a matter of great utility in the construction of machinery to determine the different degrees of cohesion possessed by different substances; or, in other words, to ascertain their *strength*. This is done by forming them into rods of equal size, and applying weights to their lower extremities sufficient to break them, by drawing them asunder. The amount of weight shows their relative degrees of strength. The following table gives the weights required to break the different substances, *each being formed into a rod one quarter of an inch square:*

Woods.

Ash, toughest	1000	lbs.
Beech	718	"
Box	1250	"
Cedar	712	"
Chestnut	656	"
Elm	837	"
Locust	1280	"
Maple	656	"
Oak, white	718	"
Pine, white	550	"
" pitch	750	"
Poplar	437	"
Walnut	487	"

Metals.

Steel, best	9370	lbs.
" soft	7500	"
Iron, wire	6440	"
" best bar	4690	"
" common bar	3750	"
" inferior bar	1880	"
" cast	1150 to 3100	"
Copper, wire	3800	"
" cast	2030	"
Brass	2800	"
Platina wire	3300	"
Silver, cast	2500	"
Gold, cast	1250	"
Tin	310	"
Zinc, cast	160	"
" sheet	1000	"
Lead, cast	55	"
" milled	207	"

From these tables we may ascertain the strength of chains, rods, etc., when made of different metals, and of timbers, bars, levers, swing-trees, and farm implements, when made of woods. Wood which will bear a very heavy weight for a minute or two, will break with two-thirds of the weight when left upon it for a long time. This explains the reason that store-house and barn timbers sometimes give way under heavy loads of grain, which have appeared at first to stand with firmness.

Although the preceding table gives the strength of wood drawn *lengthwise*, yet the comparative results are not greatly different when the force is applied in a *transverse* or *side* direction, so as to break in the usual way.

The following table shows the results of several experiments with pieces of wood one foot in length, one inch square, with the weight suspended from one end, breaking them sidewise.

White oak, seasoned, broke with			240 lbs.
Chestnut,	"	"	170 "
White pine,	"	"	135 "
Yellow pine,	"	"	150 "
Ash,	"	"	175 "
Hickory,	"	"	270 "

A rod of good iron is about ten times as strong as the best hemp rope of the same size. The best iron wire is nearly twenty times as strong as a hemp cord. Hence the enormous strength of the wire cables, several inches in diameter, which are employed for the support of suspension bridges.

A rope one inch in diameter will bear about 5000 lbs., but in practice should not be subjected to more than half this strain, or about one ton. The strength increases or diminishes according to the size of the cross-section of the rope; thus a cord half an inch in diameter will support one quarter as much as an inch, and a quarter-inch cord a sixteenth as much. A knowledge of the strength of ropes, as used by farmers in windlasses, pulleys, drawing loads, etc., would sometimes prevent serious accidents. The following table may therefore be useful:

Diameter of rope or cord in inches.	Pounds borne with safety.	Breaking weight.
One-eighth	31 lbs.	78 lbs.
One-fourth	125 "	314 "
One-half	500 "	1250 "
One	2000 "	5000 "
One and a quarter	3000 "	7500 "
One and a half	4500 "	12,500 "

These results will vary about one-fourth with the quality of common hemp. Manilla is about one-half as strong as the best hemp. The latter stretches one-fifth to one-seventh before breaking.

Wood is about seven to twenty times stronger when taken lengthwise with the fibres than when a side force is exerted, so as to *split* it. The splitting of timber or wood for fuel is, however, accomplished with a comparatively small power by the use of wedges, the force of heavy blows, and the leverage of the two parts.

The attraction of cohesion is very weak in liquids; it is sufficient, however, to give a round or spherical shape to very small portions or single drops, and to furnish a beautiful illustration, on a minute scale, of the same principle which gives a rounded form to the surface of the sea. In one case, cohesion, by drawing toward a common centre, forms the minute globule of dew upon the blade of grass; in the other, gravitation, acting in like manner, but at vast distances, gives the mighty rotundity to the rolling waters of the ocean.

CAPILLARY ATTRACTION.

Capillary attraction is a species of cohesion; it takes place only between solids and liquids. It is this which holds the moisture on the surface of a wet body, and which prevents the water from running instantly out of a wet cloth or sponge. By touching the lower extremity of a lump of sugar to the surface of water in a vessel, capillary attraction will cause the water to rise among its granules and moisten the whole lump. It may be very distinctly shown by placing the end of a fine glass tube into water; the water will rise in it above the level of the surrounding surface. If the bore of the tube be the twelfth of an inch in diameter (*a*, fig. 15,) it will rise a quarter of an inch; if but the twenty-fifth of an inch in bore, as *b*, it will rise half an inch; but if only a fiftieth of an inch, the water will rise an inch. This ascent of the liquid is caused by the attraction of the inner surface of the tube, until the weight of the column becomes equal to the force of the attraction. Capillary attraction may be also exhibited by

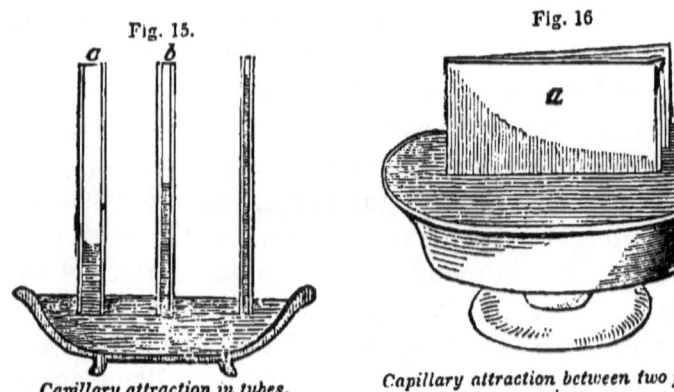

Fig. 15. *Capillary attraction in tubes.*

Fig. 16. *Capillary attraction between two panes of glass.*

two small plates of glass, placed with their edges in water, in contact on one side, and a little open at the other side, as in fig. 16. As the faces of the plates approach each other, the water rises higher, forming the curve, *a*.

Capillary attraction performs many important offices in nature. The moisture of the soil depends greatly upon its action. If the soil is composed of coarse sand or gravel, the interstices are large, and, like the larger glass tube, will not retain the rain which falls upon it. Such soils are, therefore, easily worked in wet weather, but become too dry in seasons of drought; but when the texture is finer, and especially if a due proportion of clay be mixed with the sand, the interstices become exceedingly small, and retain a full sufficiency of moisture. If, however, there is too much clay, the soil is apt to become close and compact, and the water can not enter until it is broken up or pulverized. It is for this reason that subsoil plowing becomes so eminently beneficial, by deepening the mellow portion, and thus affording a larger reservoir, which acts like a sponge in holding the excess of falling rains, until wanted in the dry season. For the same reason, a well-cultivated soil is found to preserve its moisture much better during the heat of summer than a hardened and neglected surface.

If capillary attraction should cease to exist, the earth would soon become a barren and uninhabitable waste. The moisture of rains could not be retained by the particles of the soil, but would immediately sink far down into the earth, leaving the surface at all times as dry and unproductive as a desert; vegetation would cease; brooks and rivers would lose the gradual supplies which the earth affords them through this influence, and become dried up; and all plants and all animals die for want of drink and nourishment. Thus the very existence of the whole human race evidently depends on a law, apparently insignificant to the unthinking, but pointing the observing mind to a striking proof of the creative design which planned all the works of nature, and fitted them with the utmost exactness for the life and comfort of man.

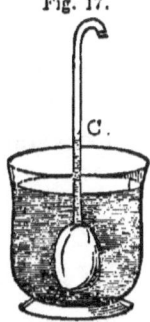

Fig. 17.

Apparatus explaining the rising of sap.

ASCENT OF SAP.

The following interesting experiments serve to explain the cause of the ascent of sap in plants and trees:

Take a small bladder, or bag made of any similar substance, and fasten it tightly on a tube open at both ends (fig. 17); then fill them with alcohol up to the point C, and immerse the bladder into a vessel of water. The alcohol will immediately rise slowly in the tube, and if not more than two or three feet high, will run over the top. This is owing to the capillary attraction in the minute pores of the bladder, drawing the water within it faster than the same attraction draws the alcohol outward. One liquid will thus intrude itself into another with great force. A bladder filled with alcohol, with its neck tightly tied, will soon burst if plunged under water. If a bladder is filled with gum-water, and then immersed as before, the water will find its way within against a very heavy pressure.

In this manner sap ascends through the minute tubes in the body of trees. The sap is thickened like gum-water when it reaches the leaves, and a fresh supply, therefore, enters through the pores in the spongelets of the roots by capillary attraction, and, rising through the stem, keeps up a constant supply for the wants of the growing tree.

CENTRE OF GRAVITY.

The *centre of gravity* is that point in every hard substance or body, on every side of which the different parts exactly balance each other. If the body be a globe or round ball, the centre of gravity will be exactly at the centre of the globe; if it be a rod of equal size, it will be at the middle of the rod. If a stone or any other substance rest on a point directly under the centre of gravity, it will remain balanced on this point; but if the point be not under the centre of gravity, the stone will fall toward the heaviest side.

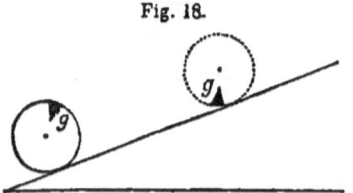

Fig. 18.

Some curious experiments are performed by an ingenious management of the centre of gravity. A light cylinder of cork or pasteboard contains a concealed piece of lead, *g* (fig. 18). The lead, being heavier than the rest, will cause the cylinder to roll up an inclined plane, when placed as shown by the lower figure on the preceding engraving, until it makes half a revolution and reaches the place of the upper figure, when it will remain stationary. If a curved body, as shown in fig. 19, be loaded heavily at its ends, it will rest on the stand, and present a singular appearance by not falling, the centre of gravity lying between the two heavy portions on the end of the stand. A light stick of some length may be made to stand on the end of the finger, by sticking in two penknives, so as to bring the centre of gravity as low as the finger-end (fig. 20).

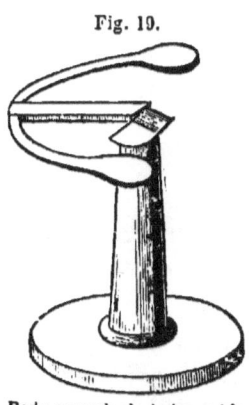

Fig. 19.

Body singularly balanced by lead knobs.

If any body, of whatever shape, be suspended by a hook or loop at its top, it will necessarily hang so that the centre of gravity shall be directly under the hook. In this way the centre in any substance, no matter how irregular its shape may be, is ascertained. Suppose, for instance, we have the irregular plate or board shown in the annexed figure (fig. 21): first hang it by the hook *a*, and the centre of gravity will be somewhere in

Fig. 20.

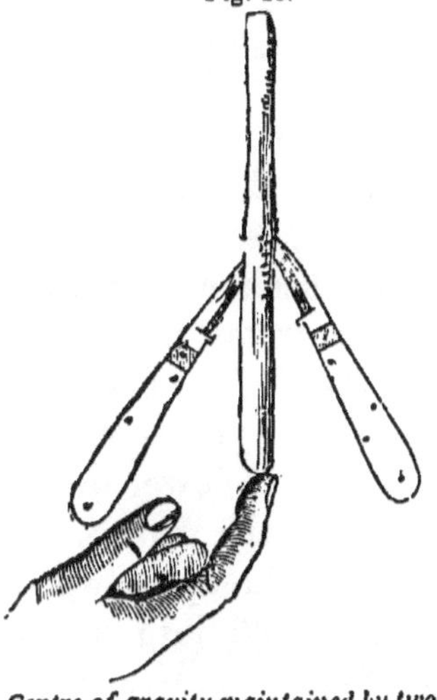

Centre of gravity maintained by two penknives.

Fig. 21.

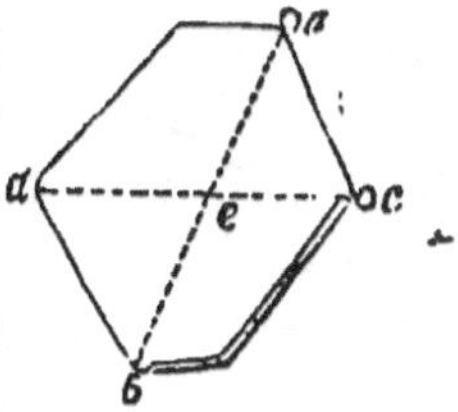

the dotted line *a b*. Then hang it by the hook *c*, and it will be somewhere in the line *c d*. Now the point *e*, where they cross each other, is the only point in both, consequently this is the centre sought. If the mass or body, instead of being flat like a board, be shapeless like a stone or lump of chalk, holes bored from different suspending points directly downward will all cross each other exactly at the centre of gravity.

LINE OF DIRECTION.

An imaginary line from the centre of gravity perpendicularly downward to where the body rests is called the *line of direction.*

Now in any solid body whatever, whether it be a wall, a stack of grain, or a loaded wagon, the line of direction must fall within the base or part resting upon the ground, or it will immediately be thrown over by its own weight. A heavily and evenly loaded wagon on a level road will be perfectly safe, because the line of direction falls equally between the wheels, as shown in fig. 22, by the dotted line, *c*, being the centre. But if it pass a steep sidehill road, throwing this line outside the wheels, as in fig. 23, it must be instantly overturned. If, however, instead of the high load represented in the figure, it be some very heavy material, as brick or sand, so as not to be higher than the square box, the centre will be much lower down, or at *b*, and thus, the line falling within the wheels, the load will be safe from upsetting, unless the upper wheel pass over a stone, or the lower wheel sink into a rut. The centre of gravity of a large load may be nearly ascertained by measuring with a rod; and it may sometimes happen that by measuring the sideling slope of a road, all of which may be done in a few minutes, a teamster may save himself from a comfortless upsetting, and perhaps

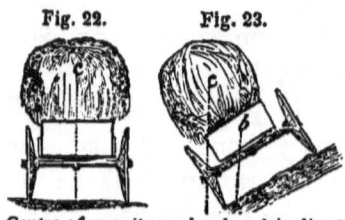

Fig. 22.   Fig. 23.

Centre of gravity on level and inclined roads.

heavy loss. Again, a load may be temporarily placed so much toward one side, while passing a sideling road, as to throw the line of direction considerably more up hill than usual, and save the load, which may be adjusted again as soon as the dangerous point is passed. This principle also shows the reason why it is safer to place only light bundles of merchandise on the top of a stage-coach, while all heavier articles are to be down near the wheels; and why a sleigh will be less likely to upset in a snow-drift, if all the passengers will sit or lie on the bottom.

When it becomes necessary to build very large loads of hay, straw, wool, or other light substances, the "reach," or the long connecting-bar of the wagon, must be made longer, so as to increase the length of the load; for, by doubling the length, two tons may be piled upon the wagon with as much security from upsetting as one ton only on a short wagon.

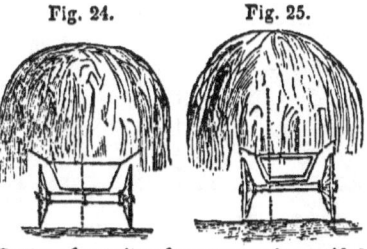

Centre of gravity of an even and one-sided load.

Where, however, a high load can not be avoided, great care must be taken to have it evenly placed. If, for instance, the load of hay represented by fig. 24 be skillfully built, the line of direction will fall equally distant within each wheel; but a slight misplacement, as in fig. 25, will so alter this line as to render it dangerous to drive except on a very even road.

Thus every one who drives a wagon should understand the laws of nature sufficiently to know how to arrange the load he carries. It is true that experience and good judgment alone will be sufficient in many cases; but no person can fail to judge better, with the reasons clearly, accurately, distinctly before his eyes, than by loose conjecture and random guessing.

Every farmer who erects a wall or building, every teamster who drives a heavy load, or even he who only carries a heavy weight upon his shoulder, may learn something useful by understanding the laws of gravity.

It is familiar to every one, that a body resting upon a broad base is more difficult to upset than when the base is narrow. For instance, the square block (fig. 26) is less easily thrown over than the tall and narrow block of equal weight, because, in turning the square block over its lower edge, the centre of gravity must be lifted up considerably in the curve shown by the dotted line c; but with a tall, narrow block, this curve being almost on a level, very little lifting is required. Hence the reason that a high load on a wagon is so much more easily overturned than a low one.

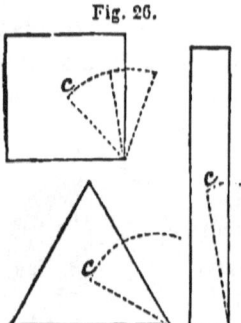

Fig. 26.

Of all forms, a pyramid stands the most firmly on its base. The centre of gravity, c (fig. 26), being so near the broad bottom, it must be elevated in a very steep curve to throw the line of direction beyond the base. For this reason, a stone wall, or the dam for a stream, will stand better when broad at bottom and tapering to a narrow top than if of equal thickness throughout.

When a globe or round ball is placed upon a smooth floor, it rests on a single point. If the floor be *level*, the line of direction will fall exactly at this resting-point (fig. 27). To move the ball, the centre will move precisely on a level, without being raised at all. This is the reason that a ball, a cylinder, or a wheel is rolled forward so much more easily than a flat-sided or irregular body. In all these cases, the line of direction, although constantly

Fig. 27.

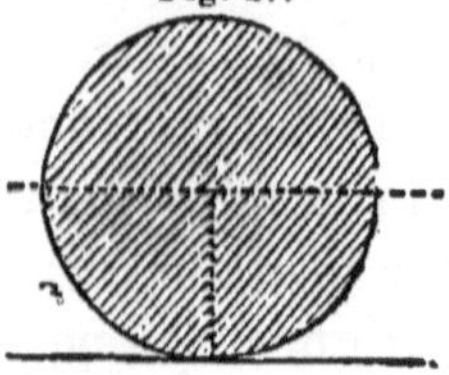

CENTRE OF GRAVITY.—EXAMPLES.   39

changing its place, still continues to fall on the very point on which the round body rests.

But if the level floor is exchanged for a slope or inclined plane (fig. 28), the line of direction no longer falls at the touching-point, but on the side from it downward; the ball will therefore, by its mere weight, commence rolling, and continue to do so until it reaches the bottom of the slope.

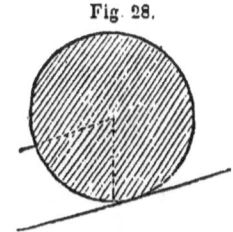

Fig. 28.

Wheel-carriages owe their comparative ease of draught to the fact that the centre of gravity in the load is moved forward by the rolling of the wheels, on a level, or parallel with the surface of the road, just in the same way that the round ball rolls so easily. Each wheel supporting its part of the load at the hub, the same rule applies to each as to a ball or cylinder alone. Hence, on a level road, the line of direction falls precisely where the wheels rest on the ground, but if the road ascend or descend, it falls elsewhere; this explains the reason why it will run by its own weight down a slope.

Whenever a stone or other obstruction occurs in a road, it becomes requisite to raise the centre by the force of the team and by means of oblique motion, so as to throw the wheel over it, as shown by fig. 29. One of the reasons thus becomes very plain why a large wheel will run with more ease on a rough road than a smaller one; the larger one mounting any stone or obstruction without lifting the load so much out of a level or direct line, as shown by the dotted lines in the annexed figures, (figs. 29

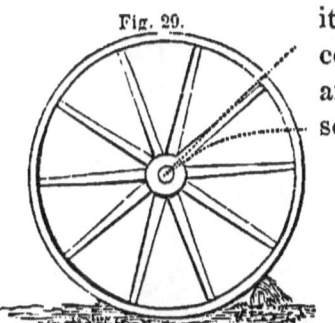

Fig. 29.

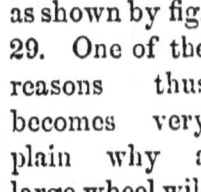

Fig. 30.

and 30). Another reason is, the large wheel does not sink into the smaller cavities in the road.

A self-supporting fruit-ladder (fig. 31) (the centre of gravity, when in use, being at or near the top) must have its legs more widely spread, to be secure from falling, than if the centre were lower down. Hence such a position, as in fig. 32, would be unsafe.

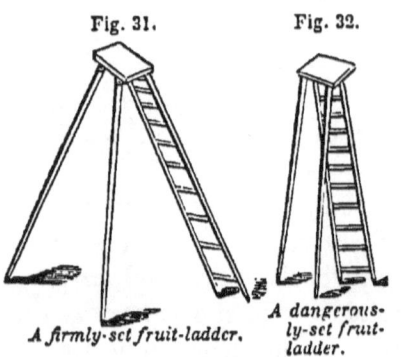

Fig. 31. Fig. 32.

A firmly-set fruit-ladder. A dangerously-set fruit-ladder.

The support of the human body, in standing and walking, exhibits some interesting examples in relation to this subject. A child can not learn to walk until he acquires skill enough to keep his feet always in the line of direction. When he fails to do this, he topples over toward the side where the line falls outside his feet. A man standing with his heels touching the wash-board of a room can not possibly stoop over without falling, because, when he bends, the line of direction falls forward of his toes, the wall against which he stands preventing the movement of his body backward to preserve the balance.

Fig. 33.

In walking, the centre rises and falls slightly at each step, as shown by the waved line in fig. 33. If it were not for the bending of the knee-joints, this exercise would be much more laborious, as it would then become needful to throw the centre into an upward curve at every step. For this reason, a wooden leg is more imperfect than the natural one (fig. 34). Hence the reason why walking on crutches is laborious and fatiguing, because at every on-

Fig. 34.

ward step the body must be thrown upward in a curve, like a wagon mounting repeated obstructions.

When a load is carried on the shoulder, it should be so placed that the line of direction may pass directly through the shoulder or back down to the feet, fig. 35. An unskillful person will sometimes place a bag of grain as shown in fig. 36. The line falling outside his feet, he is compelled to draw downward with great force on the other end of the bag. A man who carries a heavy pole on his shoulder should see that the centre is directly over his shoulder, otherwise he will be compelled to bear down upon the lighter end, and thus add in an equal degree to the weight upon his body.

Fig. 35.   Fig. 36.

If an elliptical or oval body, fig. 37, rest upon its side a, rolling it in either direction elevates the centre, c, because it is nearest the side on which the body rests. If, when raised, it be suffered to fall, its momentum carries it beyond the point of rest, and thus it continues rocking until the force is spent. The course of the centre during these motions is shown by the curved dotted line, c. If it be placed upon end, as in fig. 38, then any motion toward either side brings the centre of gravity nearer the touching-point, that is, causes it to descend, and the body consequently falls over on its side. This may be easily illustrated with an egg, which will lie at rest upon its side, but falls when set on either end.

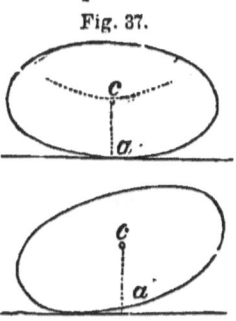

Fig. 37.

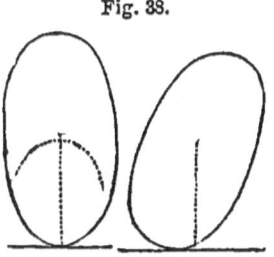

Fig. 38.

The rockers of chairs, cradles, and cribs, are formed on the principle just explained. If so made that the centre of gravity

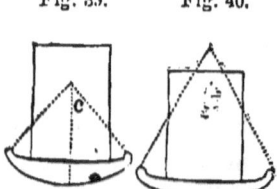

Fig. 39.   Fig. 40.

of the chair or cradle is nearer the middle of the rocker than to the ends, the rocking motion will take place; and when the distance from the centre of gravity to the ends of the rockers is but little greater than the distance to the middle, *c*, as in fig. 39, the motion will be slow and gentle; but if this difference be greater, as in fig. 40, it will be rapid. When the centre is high, the rockers must have less curvature than where it is low and near the floor. If the centre of gravity be nearer the ends than to the middle, the chair will immediately be overturned. This principle should be well understood in the construction of every thing which moves by rocking.

CHAPTER IV.

SIMPLE MACHINES, OR MECHANICAL POWERS.

ADVANTAGES OF MACHINES.

The moving forces which are applied to various useful purposes commonly require some change in velocity, direction, or mode of acting, before they accomplish the desired end. For example, a running stream of water has a motion in one direction only; by the use of machinery, we change this to an alternating motion, as in the saw of the saw-mill, or to a rotatory or whirling motion, as in the stones of a grist-mill. The direct or straightforward power of a yoke of oxen is made, by the employment of the plow, to produce a side-motion to the sod, as well as to turn it through half a circle. The thrashing-machine converts the slowly-acting pace of horses to the swift hum of the spiked cylinder.

Any instrument used for thus changing or modifying motion is called *a machine*, whether it be simple or complex in its structure. Thus even a *crow-bar*, used in lifting stones from the earth, by diminishing the motion given by the hand and increasing its power, may be strictly termed a machine; while a *harrow*, which neither alters the course nor changes the velocity of the force applied, may with more propriety be regarded as simply an implement or tool. In common language, however, these distinctions are not accurately observed, and a machine is usually considered to be any instrument consisting of different moving parts.

All machines, however complex, may be resolved into two simple parts, or powers. These are,

1. THE LEVER;
2. THE INCLINED PLANE.

The *wheel and axle*, and the *pulley* are modified applications of the LEVER; and the *wedge* and the *screw* of the INCLINED PLANE, as will be shown on the following pages. These six are usually termed the *mechanical powers*. As they really do not possess any *power* in themselves, but only regulate power, the term "simple machines" may be regarded as most correct.

THE LAW OF VIRTUAL VELOCITIES.

Before proceeding to the simple machines, it may be well to explain a very important truth, which should be considered as lying at the foundation of all mechanical philosophy, and which renders plain and simple many things which would otherwise seem strange or contradictory. This is, that the force required to lift any given body is always in proportion to the weight of that body, taken together with the height to be raised. For instance, it requires twice the force to raise two pounds as to raise one pound, three times the force to raise three pounds, and so forth. Also, twice as great a force is needed to elevate any weight two feet as one foot, or three times as great for three feet, and so on. Again, combining these together, four times as great a force is required to raise two pounds to a height of two feet as to raise one pound only one foot; eight times as great for four feet, and so on. This holds true, no matter by what kind of machinery it is accomplished. Now this may all seem very simple, but it serves to explain many difficult questions in relation to the real power possessed by all machines.

Take another example. Suppose that one wishes to raise a weight of 1000 pounds to a height of one foot. If his strength is equal to only 100 pounds, the weight would be ten times too heavy for him. He might, therefore, divide it into ten equal parts of 100 pounds each. Raising each part separately the required height of one foot would be the same as raising one of them ten feet high. The weight is lessened ten times, but the distance is increased ten times. But there are some bodies, as, for example, blocks of stone or sticks of timber, which can not well be divided into parts in actual practice. He therefore resorts to a machine or mechanical power, through which the same result is accomplished by raising the whole weight in one mass with his single strength; but in this case as well as the other, the moving force which he applies must pass through ten times the space of the weight. We arrive, therefore, at the general rule, that the distance moved by the weight is as much less than that moved by the power as the power is less than the weight. This rule is termed by some writers the "*rule of virtual velocities,*" *virtual* meaning not apparent or actual, but according to the real effect, because the increase in the velocity of the power makes up for increase in the size of weight. This rule will be better understood after considering its application to the different simple machines.

The simplest of all machines is the *lever*. It consists of a rod or bar, one end resting upon a prop or *fulcrum*, F (fig. 41), near which is the weight, W, moved by the hand at P. The stone may weigh 1000 pounds; yet, if it is ten times as near the fulcrum as the man's hand is, a force of 100 pounds will lift it; but it will be moved only a tenth part as high as the hand has been moved, as shown

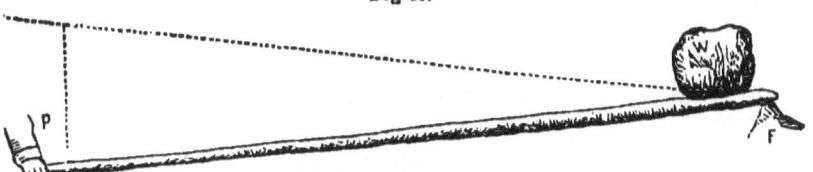

Fig 41.

Lever of the second kind.

by the dotted lines. By placing the stone still nearer the fulcrum, still less will be the power required to raise it, but then the distance elevated would be also still less. By sufficiently increasing the disproportion between the two parts of the lever, the strength of a child merely might be made to move more than many horses could draw.

These performances of the lever often excite astonishment at what appears to be out of the common course of things; yet, when examined by the principles of mechanics, instead of appearing matters of astonishment, they are found to be only the natural and necessary results of the laws of force. In the case of the lever just described, it is often incorrectly supposed that the power itself sustains the weight. But this is not the case; nearly the whole of it rests upon the fulcrum. We often see proofs of this error in common practice, where fulcrums or props entirely insufficient to uphold the enormous weight to be raised are attempted to be used. If the weight, for instance, be ten times as near the fulcrum as to the power, then nine-tenths of the weight rests upon the fulcrum, and the remaining tenth only is sustained by the lifting power. The lever only allows the power to expend itself through a longer distance, and thus, by concentrating itself at the weight, to elevate the latter through the shorter distance, according to the rule of virtual velocities already explained.

The fulcrum may be placed between the weight and the power, as in fig. 42, or the power may be placed between the fulcrum and the weight, as in fig. 43, the same principle of virtual velocities applying in all cases.

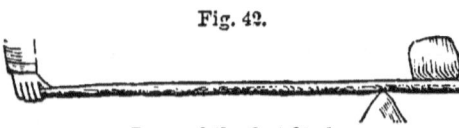

Lever of the first kind.

Where the fulcrum is between the power and the weight, as in fig. 42, it is called a *lever of the first kind*.

Where the weight is between the fulcrum and the power, as in fig. 41, it constitutes a *lever of the second kind*.

Where the power is between the fulcrum and the weight, as in fig. 43, it is termed a *lever of the third kind*.

1. Many examples occur in practice of levers of the first kind. A crow-bar, used to raise stones from the earth, is an instance of this sort; so is a handspike of any kind used in the same way. A hammer for drawing a nail

Lever of the third kind.

operates as a lever of the first kind, the heel being the fulcrum, the nail the weight, and the hand the power; the distance through which the handle passes being several times greater than that of the claws, the force exerted on the nail is increased in like proportion. A pair of scissors consists of two levers, the rivet being the fulcrum; and in using them, as every one has observed, a greater cutting force is exerted near the rivets than toward the points.

This is owing to the power being expended through a greater distance near the points, according to the rule already explained. Pincers, nippers, and other similar instruments are also double levers of the first kind.

A common steelyard is another example, the sliding weight becoming gradually more effective as it is moved further from the fulcrum or hook supporting the instrument. The brake or handle of a pump is a lever of this class, the pump-rod and piston being the weight.

The common balance is still another, the two arms being exactly equal, so that one weight will always balance the other, and on this its usefulness and accuracy entirely depend. The most sensitive balances have light beams with long arms, and the turning-point of hardened steel or agate, in the form of a thin wedge, on which the balance turns almost without friction. Small balances have been so skillfully constructed as to turn with one-thousandth part of a grain.

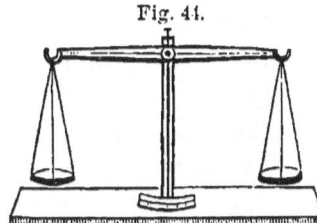

Fig. 41.

2. Levers of the second kind are less numerous, but not uncommon. A handspike used for rolling a log is an example. A wheel-barrow is a lever of the second kind, the fulcrum being the point where the wheel rests on the ground, and the weight the centre of gravity of the load. Hence, less exertion of strength is required in the arm when the load is placed near the wheel, except where the ground is soft or muddy, when it is found advantageous to place the load so that the arm shall sustain a considerable portion, to prevent the wheel sinking into the soil. A two-wheeled cart is a similar example; and, for the same reason, when the ground is soft, the load should be placed forward toward the horse or oxen; on the other hand, on a smooth and hard, or on a plank road, the load should be

more nearly balanced. An observance of this rule would often save a great deal of needless waste of strength.

A sack-barrow, used in barns and mills for conveying heavy bags of grain from one part of the floor to another,

Fig. 45.

Sack-barrow.

and in warehouses for boxes, is a lever nearly intermediate between the first and second kind, the weight usually resting very nearly over the fulcrum or wheels. When the bag of grain is thrown forward of the wheels, it becomes a lever of the first kind; when back of the wheels, it is a lever of the second kind. As it is used only on hard and smooth floors, and not, like the wheel-barrow, on soft earth, the more nearly the load is placed directly over the wheels, the more easily they will run.

3. In a lever of the third kind, the weight being further from the fulcrum than the power, it is only used where great power is of secondary importance when compared with rapidity and dispatch. A hand-hoe is of this class, the left hand acting as the fulcrum, the right hand as the power, and the resistance overcome by the blade of the hoe as the weight. A hand-rake is similar, as well as a fork used for pitching hay. Tongs are double levers of this kind, as also the shears used in shearing sheep. The limbs of animals, generally, are levers of the third

kind. The joint of the bone is the fulcrum; the strong muscle or tendon attached to the bone near the joint is the power; and the weight of the limb, with whatever resistance it overcomes, is the weight. A great advantage results from this contrivance, because a slight contraction of the muscle gives a swift motion to the limb, so important in walking and running, and in the use of the arms.

ESTIMATING THE POWER OF LEVERS.

The power of any lever is easily calculated by measuring the length of its two arms, that is, the two parts into which it is divided by the weight, fulcrum, and power. In a lever of the first kind, if the weight and power be equally distant from the fulcrum, they will move through equal distances, and nothing will be gained; that is, a power of 100 pounds will lift a weight of 100 pounds only. If the power be twice as far as the weight, its force will be doubled; if three times, it will be tripled; and so forth. In a lever of the second kind, if the weight be equidistant between the fulcrum and power, the power will move through twice the distance of the weight, and the power of the instrument will therefore be doubled; if twice as far, it will be tripled, and so on, as shown in the annexed figures. The same mode of reasoning will explain precisely to what extent the force is diminished in levers of the third kind.

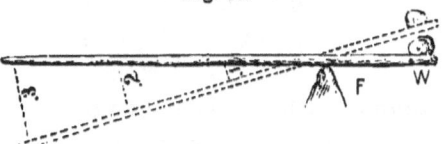

Fig. 46.
Lever of the first kind.

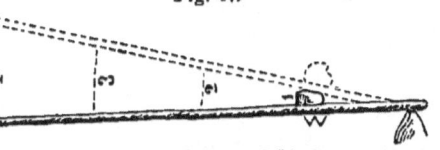

Fig. 47.
Lever of the second kind.

These rules will show in what manner a load borne on a pole is to be placed between two persons carrying it,

If equidistant between them, each will sustain a like portion. If the load be twice as near to one as to the other, the shorter end will receive double the weight of the longer. For the same reason, when three horses are worked abreast, the two horses placed together should have only half the length of arm of the main whiffle-tree as the single horse, fig. 48. The farmer who has a team of two horses unlike in strength, may thus easily know how to adjust the arms of the whiffle-tree so as to correspond with the strength of each. If, for instance, one of the horses possesses a strength as much greater than the other as four is to three, then the weaker horse should be attached to the arm of the whiffle-tree made as much longer than the other arm as four is to three.

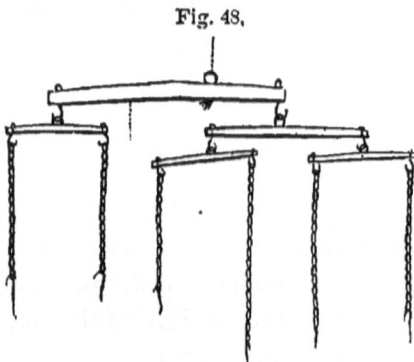

Fig. 48.

In all the preceding estimates, the influence of the weight of the lever has not been taken into consideration. In a lever of the first kind, if the thickness of the two arms be so adjusted that it will remain balanced on the fulcrum, its weight will have no other effect than to increase the pressure on the fulcrum; but if it be of equal size throughout, its longer arm, being the heavier, will add to its power. The amount thus added will be equal to the excess in the weight of this arm, applied so far along as the centre of gravity of this excess. If, for example, a piece of scantling twelve feet long, *a b*, fig. 49,

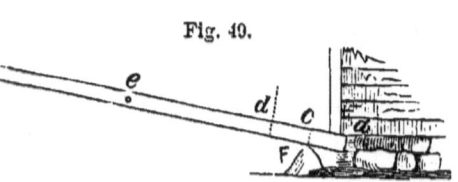

Fig. 49.

be used as a lever to lift the corner of a building, then the two portions, *a c*, *c d*, will mutually balance each other. If these be each a foot in length, the weight of ten feet will be left to bear down the lever. The centre of gravity of this portion will be at *e*, six feet from the fulcrum, and it will consequently exert a force under the building equal to six times its own weight. If the scantling weigh five pounds to the foot, or fifty pounds for the excess, this force will be equal to three hundred pounds.

In the lever of the second kind, its weight operates *against* the moving power. If it be of equal size throughout, this will be equal to just one-half the weight of the lever, the other half being supported by the fulcrum.

With the lever of the third kind, the rule applied to the first must be exactly reversed.

COMBINATION OF LEVERS.

A great power may be attained without the inconvenience of resorting to a very long lever, by means of a *combination of levers*. In fig. 50, the small weight P, acting as a moving power, exerts a three-fold force on the next lever; this, in its turn, acts in the same degree on the third, which again increases the power three times. Consequently, the moving power, P, acts upon the weight, W, in a twenty-seven-fold degree, the former passing through a space twenty-seven times as great as the latter.

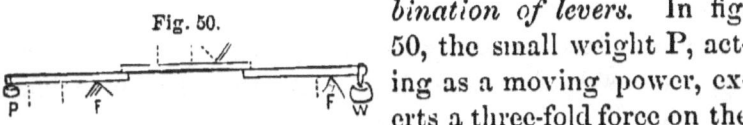
Fig. 50.

A combination of levers like this is employed in self-regulating stoves. It is in this case, however, used to multiply instead of to diminish motion. The expansion of a metallic rod by heat the hundredth part of an inch acts on a set of iron levers, and the motion is increased, by the time it reaches the draught-valve, to about one hundred times.

A more compact arrangement of compound levers is shown in fig. 51, where the power, P, acts on the lever A, exerting a force on the lever B five times as great as the power. B acts on the lever C with a force increased three times, and this, again, on the weight, W, with a four-fold force. Multiplying 5, 3, and 4 together, the product is 60; hence a force of one pound at P will support 60 pounds at W. By graduating (or marking into notches) the lever C, so that the distance is measured as the weight is moved along it, a compact and powerful steelyard for weighing is formed.

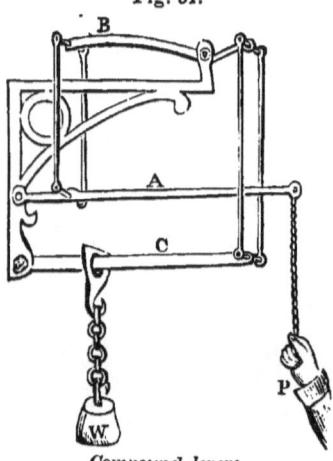

Fig. 51.

Compound levers.

WEIGHING MACHINE.

A valuable combination of levers is made in the construction of the *weighing machine*, used for weighing cattle, wagons loaded with hay, and other heavy articles.

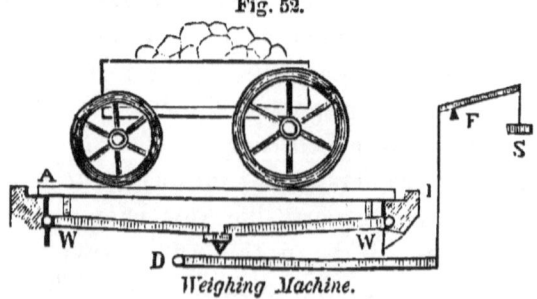

Fig. 52.

Weighing Machine.

The wagon rests on the platform A (fig. 52,) and this platform rests on two levers at W, W, which presses their other ends both on a central point, and this again bears on the lever D, the other end of which is connected by means of an upright rod with the steelyard at F.

There are two important points gained in this combination. In the first place, the levers multiply the power so much that a few pounds' weight will balance a heavy load of hay weighing a ton or more; and, in the next, the load resting on both the levers, communicates the same force of weight to the central point, from whatever part of the platform it happens to stand on: for if it presses hardest on one lever, it bears lighter, at a corresponding rate, on the other. In practice, there are always two pairs, or *four* levers, which proceed from each

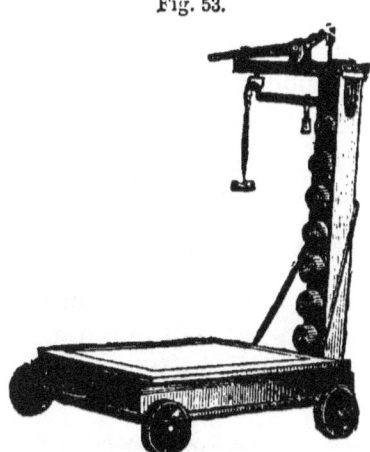

Fig. 53.

Portable Platform Scale.

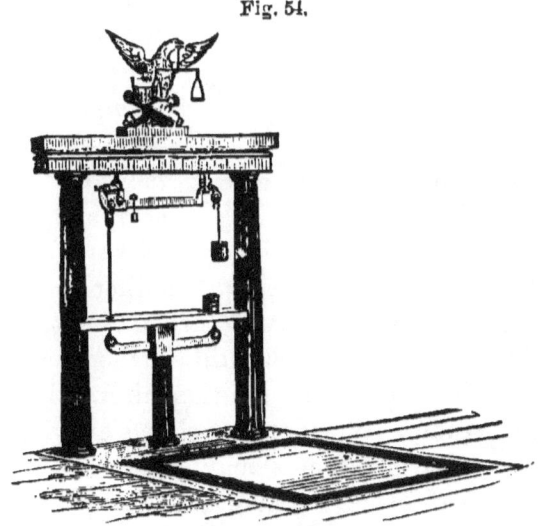

Fig. 54.

Large Platform Scale.

corner of the platform, and rest on one point at the centre. We have taken the two only, to simplify the explanation.

A powerful stump-extracting machine, allowing a succession of efforts in the use of the lever, is exhibited by fig. 55. The lever, *a*, should be a strong stick of timber, furnished with three massive iron hooks, secured by bolts passing through, as represented in the figure. Small or truck wheels are placed at each end of the lever, merely for the purpose of moving it easily over the ground. The stump, *b*, used as a fulcrum, has the chain passing round near its base, while another chain passes over the top of the stump, *c*, to be torn out. A horse is attached to the lever at *d*, and, moving to *e*, draws the other end of the lever backward, and loosens the stump; while in this position, another chain is made to connect *g* to *h*, and the horse is turned about, and draws the lever backward to *i*, which still further increases the loosening; a few repetitions of this alternating process tear out the stump. Very strong chains are requisite for this purpose. Large stumps may require an additional horse or a yoke of oxen. Where the stumps are remote from each other, iron rods with hooks may be used to connect the chains.

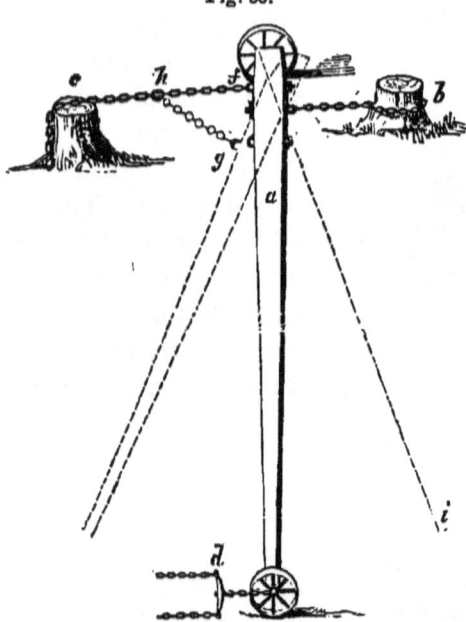

Fig. 55.

Lever Stump Machine.

The power which may be given to this and to all other

modes of using the lever, as we have already seen, depends on the difference between the lengths of its two arms. A yoke of oxen, drawing with a force of 500 pounds on the long arm of a lever 25 feet long, will exert a force on the short arm of six inches equal to 50 times 500 pounds, or 25,000 pounds, on the stump.

It was after an examination of the great power which may be given to the lever by increasing this difference that Archimedes exultingly exclaimed, "Give me but a fulcrum whereon to place my lever, and I will move the earth!" Admitting the *theoretical* truth of this exclamation, and supposing there could be a lever which he might have used for this purpose, its practical impossibility may be quickly understood by computing the whole bulk of the globe; for such is its enormous size and cubical contents, that Archimedes must have moved forward his lever with the strength of a hundred pounds and the swiftness of a cannon ball for eight hundred million years to have moved the earth the thousandth part of an inch!

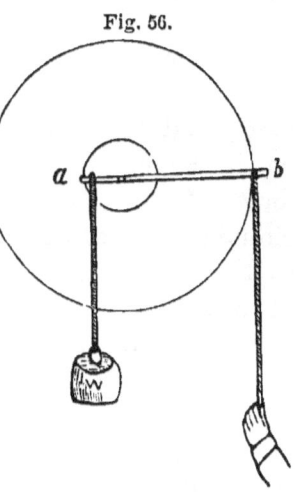

Fig. 56.

WHEEL AND AXLE.

In treating of the lever, it was shown to be capable of exerting a force through a small distance only. Hence, if a heavy body were required to be elevated to any considerable height, it would be necessary to accomplish it by a succession of efforts. This inconvenience is removed by a constant and unremitted action of the lever in the form of the *wheel* and *axle*.

Let the weight, *w* (fig. 56,) be suspended by a cord

from the end of the lever, *a b,* and a wheel attached to the lever, so that this cord may wind upon it as the weight is elevated; then let the power be applied at the other end by means of a cord, and a larger wheel be attached, so that this cord too may wind upon the larger wheel. These two wheels (fastened together so as to form one), as they are made to revolve on their axis, will now constitute, in a manner, a succession of levers, acting through an indefinite distance according to the length of the cords. The levers here successively acting are of the "first kind," and the axis of the wheel is the fulcrum. This arrangement constitutes in substance the *wheel* and *axle;* and its power, like that of the simple lever, depends on the comparative velocity of the weight and the moving force. If, for example, the larger wheel is four times the circumference of the smaller, a force of one hundred applied to the outer cord will raise a weight of four hundred pounds.

The annexed figure exhibits at one view the power exerted through the wheel and axle, where a small weight of 6 pounds will wind up (or balance) other weights separately, weighing 8, 12, or 24 pounds, as the difference increases between the size of the wheel and of the axle, according to the rule of virtual velocities already explained.

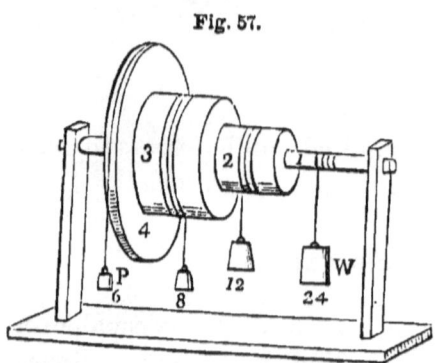

Fig. 57.

Wheel and axle, showing the heavier weight for less motion.

The thickness of the rope has not been taken into consideration. This is very small when compared with the diameter of the outer wheel, but often considerable when compared with that of the inner. To be strictly accurate, therefore, the force must be considered as acting at the centre of the rope; hence the diameter of the rope must be added to the diameter of the wheel.

There are various forms of the wheel and axle. In the common windlass, motion is given to the axle by means of a winch, which is a lever like the handle of a grindstone. The windlass used in digging wells has usually four projecting levers or arms. The wheel used in steering a vessel is furnished with pins in the circumference, to which the hand is applied in turning it. In the capstan (for weighing anchor) the axis is vertical, and horizontal levers are applied around it, so that several men may work at once. The power of all these forms is easily calculated by the rule of virtual velocities—that is, that the velocity with which the power moves is as many times greater than the velocity of the weight, as the weight exceeds the power. A simple and convenient rule for computing in numbers the power of wheel-work is the following: Multiply all the numbers together which express either the circumferences or diameters of the large wheels, and then multiply together all the numbers which express the diameters of the smaller wheels or pinions; divide the greater number by the less, and the quotient will be the power sought.

BAND AND COG WHEELS

Where great power is required, several wheels and axles may be combined in a manner corresponding with that of the compound system of levers already explained. In this case the axis of one wheel acts on the circumference of the next, producing a continued slower motion, and increasing the power in a corresponding degree.

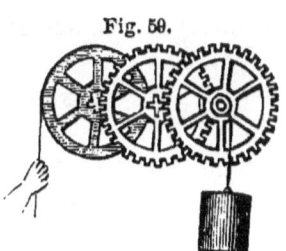

Fig. 59.

Combined cog-wheels.

The wheels are made thus to act by means of cogs or teeth,

or of bands (fig. 59). In ordinary practice, however, combined wheels are made use of to multiply motion instead of to diminish it, familiar instances of which occur in the grist-mill and thrashing machine.

In connecting a system of wheels, the cord or strap may be used where great force is not required, the friction round the circumference being sufficient to prevent slipping. Bands are chiefly useful where motion is to be transmitted to a distance; as, for example, from a horse-power without a barn to a thrashing-machine within it. Liability of sliding is sometimes useful, by preventing the machinery from breaking when a sudden obstruction occurs. Where the force is great, the necessary tension or tightness of the cord produces too great a friction at the axle. In such cases, cogs or teeth must be resorted to.

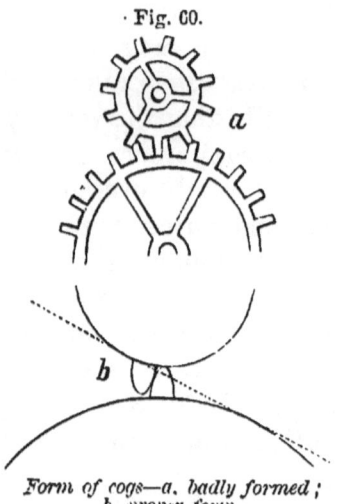

Fig. 60.

Form of cogs—a, badly formed; b, proper form.

The term *teeth* is usually applied when they are formed of the same piece as the wheel, as in the case of cast-iron wheels. *Cogs* are teeth formed separately and inserted into the wheel, as with wooden wheels. *Pinions* are the small wheels, or, more properly, teeth set on axles.

FORM OF TEETH OR COGS.

The form of the teeth has a great influence on the amount of friction among wheel-work. Badly formed teeth are represented by the wheel-work at *a*, in the annexed figure, consisting of square projecting pins. When these teeth first come into contact with each other, they act *obliquely* together, and thus a part of their force is lost, and they continue scraping together with a large amount of friction so long as they remain in contact. These effects are avoided by giving to them the curved form, represented by *b*. Here, instead of pressing each other obliquely, they act at right angles, that is, not obliquely, and instead of scraping, they *roll over* each other with ease. These curves are ascertained by mathematical calculation, which can not be here given; it may be enough to state that they should be so formed that the points in contact shall always work at right angles to each other. For ordinary practical purposes, however, they may be made as shown in the annexed figure (fig. 61), by striking circles whose diameters shall embrace just three teeth. The points of the teeth thus formed are removed, leaving a blunt extremity, according to the figure.

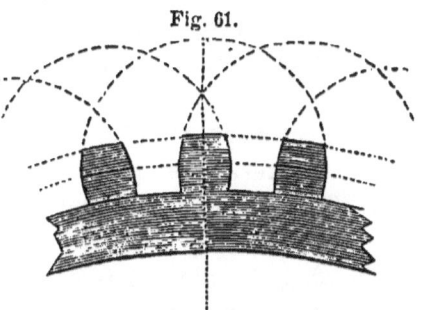

Fig. 61.

Mode of giving the best form to cogs.

There are a few other rules that should always be observed in constructing wheel-work, in order that the wheels may run easily together, without jerking or rattling, the most important of which are the following:

1. The teeth must be of uniform size and distance from each other, through the whole circumference of the wheel.

2. Any tooth must begin to act at the same instant that the preceding tooth ceases to touch its corresponding tooth on the other wheel.

3. There must be sufficient space between the teeth not only to admit those of the other wheel, but to allow a certain degree of play, which should be equal to at least one-tenth of the thickness of the teeth.

4. The pinions should not be very small, unless the

wheels they act on are quite large. In a pinion that has only eight teeth, each tooth begins to act before it reaches the line of the centres, and it is not disengaged as soon as the next one begins to act. A pinion of ten teeth will not operate perfectly if working in a wheel of less than 72 teeth. Pinions of less than six teeth should never be used.

5. To give strength to the teeth of wheels, make the wheels themselves thicker, which increases the breadth of the teeth.

6. Wheel-work is often defective when not made of uniform material, in consequence of the relative number of teeth working together not being such as to equalize the wear of all alike. If the number of teeth on a wheel is divided without a remainder by the number of the pinion, then the same teeth will repeatedly engage each other, and they will often wear unevenly. The number should be so arranged that every tooth of the pinion may work in succession into the teeth of the wheel. This is best effected by first taking a number for the wheel that will be evenly divided by the number on the pinion, and then adding *one more* tooth to the wheel. This will effect a continual change, so that no two shall be engaged with each other twice until all the rest have been gone through with. This odd tooth is called the *hunting-cog*.

Fig. 62. 
Bevel-wheels.

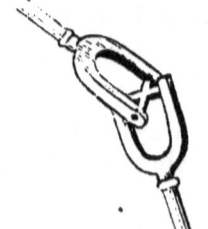

Fig. 63.
Universal Joint.

Cog-wheels are most usually made with the teeth on the outside or circumference of the wheel; these are termed *spur-wheels*.

If the teeth are set on *one side* of the wheels, they are termed *crown-wheels*. When they are made so as to work together obliquely, they are called *bevel-wheels*, as in fig. 62.

Where the obliquity is small, the motion may be communicated by means of the *universal joint*, as shown in fig. 63. This is commonly used in the thrashing-machine, where there is a slight change in the direction of motion between the horse-power and the thrasher.

THE PULLEY.

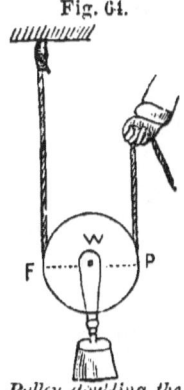

Fig. 64.

Pulley doubling the force.

Let a cord fixed at one end pass round a movable grooved wheel, and be grasped by the hand at the other end: then, in lifting any weight attached to the wheel, by drawing up the cord, the hand will move with twice the velocity of the weight. It will, therefore, exert double the degree of force. This operates precisely as a succession of levers of the second kind, the fixed cord being the fulcrum, and the cord drawn up by the hand, the power. It thus constitutes one of the simplest kinds of the pulley, fig. 64.

The wheel is called a *sheave;* the term *pulley* is applied to the block and sheave; and a combination of sheaves, blocks, and ropes is called a *tackle.*

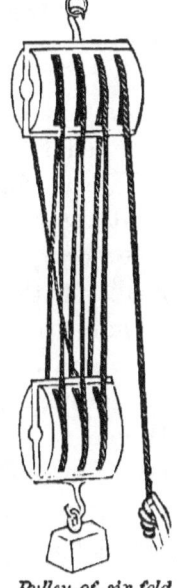

Pulley of six-fold power.

There are various combinations of single pulleys for increasing power, the most common of which, and least liable to become deranged by the cord being thrown off the wheels, is shown in fig. 65. In this and in all similarly constructed pulleys, the weight is as many times greater than the power as the number of cords which support the lower block. If there be six cords, as in the figure, the weight will be six times the power.

Where a cord is passed over a single fixed wheel, as in fig. 66, or over two or more wheels, no power is gained, the moving force being the same in velocity as the weight. Such pulleys are sometimes, however, of use by altering the *direction* of the force. The latter is applied with advantage to unloading or pitching hay by means of a horse-power, saving much time and labor, as explained on a future page.

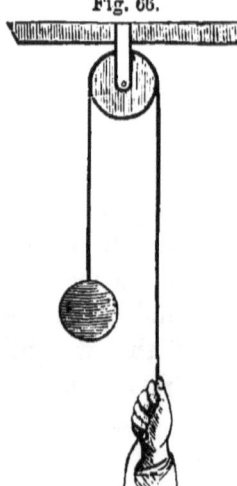

Fig. 66.

Pulley with no increase of power.

Among the many applications of the pulley, one is shown in the accompanying figure (fig. 67) representing *Packer's Stone Lifter*, for raising large boulders from the soil, weighing from one to four and five tons, and afterwards placing them in walls. It is also employed for tearing out small or partly decayed stumps.

The usefulness of the pulley depends mainly upon its lightness and portable form, and the facility with which it may be made to operate in almost any situation. Hence it is much used in building, and is extensively applied in the rigging of ships. In

Fig. 67.

Packer's Stone Lifter.

the computation of its power there is a large drawback, not taken into account in the preceding calculation, which materially lessens its advantage; this is the friction of the wheels and blocks and the stiffness of the cordage, which are often so great that two-thirds of the power is lost.

THE INCLINED PLANE.

The *inclined plane* or slope possesses a power which is estimated by the proportion which its length bears to the height. If, for example, the plane be twice as long as the perpendicular height, then in rolling the body *a* up the inclined plane (fig. 68), it will move through twice the distance required to lift it directly from *b* to *c*. Therefore only one-half the strength else required need be exerted for this purpose. The same reasoning will apply to any other proportion between the height and length; that is, the more gradual or less steep the slope becomes, the greater will be the advantage gained. A familiar example occurs in lifting a loaded barrel into a wagon: the longer the plank used in rolling it, the less is the exertion needed.

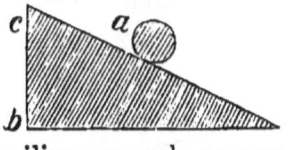

Fig. 68.

A body, in rolling freely down an inclined plane, acquires the same velocity that it would attain if dropped perpendicularly from a height equal to the perpendicular height of the plane. Thus, if an inclined plane on a plank road be 100 yards long and 16 feet high, a freely running wagon, left to descend of its own accord, will move 32 feet per second by the time it reaches the bottom, that being the velocity of a stone falling 16 feet. Or, a railcar on an inclined plane 145 feet high will attain a speed of 96 feet per second, or more than 65 miles an hour, at the foot of the plane, which is equal to the velocity of a stone falling three seconds, or 145 feet.

ASCENT IN ROADS.

All roads not perfectly level may be regarded as inclined planes. By the application of the preceding rule, we may discover precisely how much strength is lost in drawing heavy wagons up hill. If the load and wagon weigh a ton, and the road rise one foot in height to every five feet of distance, then the increased strength required to draw the load will be one-fifth of its weight, or equal to 400 pounds. If it rise only one foot in twenty, then the increase in power needed to ascend this plane will be only 100 pounds. The great importance of preserving, as nearly as practicable, a perfect level is obvious.

There are many roads made in this country, rising over and descending hills, which might be made nearly level by deviating a little to the right or to the left. Suppose, for example, that a road be required to connect the two points

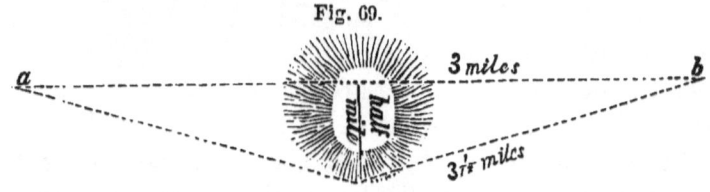

Fig. 69.

a and b (fig. 69), three miles apart, but separated by a lofty hill midway between them, and one mile in diameter. Passing half a mile on either side would entirely avoid the hill, and the road thus curved would be only one hundred and forty-eight yards, or one-twelfth of a mile longer. The same steep hill is ascended perhaps fifty to five hundred times a year by a hundred different farmers, expending an amount of strength, in the aggregate, sufficient to elevate ten thousand tons annually to this height, as a calculation will at once show—more than enough for all the increased expense of making the road level.

It is interesting and important to examine how much further it is expedient to carry a road through a circuitous level course than over a hill. To ascertain this point, we must take into view the resistance occasioned by the rough surface or soft material of the road. Roads vary greatly in this particular, but the following may be considered as about a fair average. In drawing a ton weight (including wagon) on freely running wheels, on a perfect level, the strength exerted will be found about equal to the following:

On a hard, smooth plank road..... 40 pounds.
On a good Macadam road........... 60 "
On a common good hard road....... 100 "
On a soft road about............. 200 "

Now let us compare this resistance to the resistance of drawing up hill. First, for the plank road—forty pounds is one-fiftieth of a ton; therefore a rise of one foot in fifty of length will increase the draught equal to the resistance of the road. Hence the road might be increased fifty feet in length to avoid an ascent of one foot; or, at the same rate, it might be increased a mile in length to avoid an ascent of one hundred and five feet. But in this estimate the increase in cost of making the longer road is not taken into account. If making and keeping in repair be equal to three hundred dollars yearly per mile, and one hundred teams pass over it daily, at a cost for traveling of four cents each per mile, being four dollars daily, or twelve hundred dollars per annum, then the cost of making and repair would be one quarter of the expense of traveling over it. Therefore the mile should be diminished one quarter in length to make these two sources of expense counterbalance each other. Hence a road with this amount of travel should, with a reference to public accommodation, be made three-fourths of a mile longer to avoid a hill of one hundred and five feet. This estimate applies to loaded teams only. For light carriages the advantages of the level road would not be so great. One-half to five-eighths of a mile would, therefore, be a fair estimate for all kinds of traveling taken together.

The following table shows the rise in a mile of road for different ascents:

For a rise of 1 foot in 10, the road ascends 528 feet per mile.					
do.	1 do.	13,	do.	406	do.
do.	1 do.	15,	do.	352	do.
do.	1 do.	20,	do.	264	do.
do.	1 do.	25,	do.	211	do.
do.	1 do.	30,	do.	176	do.
do.	1 do.	35,	do.	151	do.
do.	1 do.	40,	do.	132	do.
do.	1 do.	45,	do.	117	do.
do.	1 do.	50,	do.	106	do.
do.	1 do.	100,	do.	53	do.
do.	1 do.	125,	do.	42	do.

The same kind of reasoning applied to a common good road will show that it will be profitable for the public to travel about half that distance to avoid a hill of one hundred and five feet. In this case the whole yearly cost of the road, including interest on the land, and the cost of repairs, would not usually be more than a tenth part of the same cost for plank, or would not exceed thirty dollars.

On rail-roads, where the resistance is only about one-fifth part of the resistance of plank roads, the disproportion between the draught on a level and up an ascent becomes many times greater. Thus, if a single engine move three hundred and fifty tons on a level, then two engines will be required for an ascent of only twenty feet per mile, four engines for fifty feet per mile, and six engines for eighty feet per mile.

Such estimates as these merit the attention of the farmer in laying out his own private farm roads. It may be worthy of considerable effort to avoid a hill of ten or twenty feet, which must be passed over a hundred times yearly with loads of manure, grain, hay, and wood. The greatly increased resistance of soft materials, also, is too rarely taken into account. A few loads of gravel, well applied, would often prevent ten times the labor in plowing through deep ruts, to say nothing of the breaking of harness and wagons by the excessive exertions of the team.

FORM AND MATERIALS FOR ROADS.

The depth of the mud in common roads is often unnecessarily great, in consequence of heaping together with the plow and scraper the soft top-soil for the raised

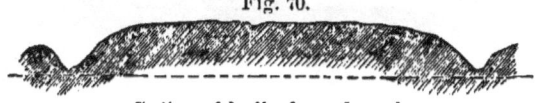
Section of badly formed road.

carriage-way. When heavy rains fall, this forms a deep bed of mud, into which the wheels work their way, and cause extreme labor to the team. A much better way is to scrape off and cart away into the fields adjoining all the soft, rich, upper surface, and then to form the harder subsoil into a slightly rounded carriage-way, with a ditch on each side. Such roads as this have a very hard and firm foundation, and they have been found not to cut up

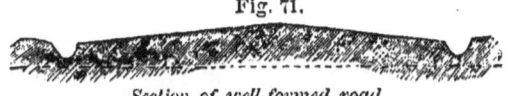
Section of well-formed road.

into ruts, nor to form much mud, even in the wettest seasons. On this hard foundation six inches of gravel will endure longer and form a better surface than twelve inches on a raised "turnpike" of soft soil and mud.

It frequently happens that the form of the surface increases the quantity of mud in a road, by not allowing the water to flow off freely. The earth is heaped up in a high ridge, but having little slope on the top (fig. 70), where the water lodges, and ruts are formed, the only dry portions being on the brink of the ditches, where the water can escape. Instead of this form, there should be a gradual inclination from the centre to the ditches, as shown in fig. 71. This inclination should not exceed 1 foot in 20. On hill-sides the slope should all be toward the higher ground, as in fig. 72.

Hard and durable roads are made on the plan of Telford. Their foundation is rounded stones, placed upright, with the smaller or sharper ends upward. The smaller stones

Section of road for hill-sides.

are placed near the sides, and the larger at the centre, thus giving to the road a convex form. The spaces are then filled in with small broken stone, and the whole covered with the same material or with gravel. The pressure of wagons crowds it compactly between the stones, and forms a very hard mass.

IMPORTANCE OF GOOD ROADS.

The principles of road-making should be better understood by the community at large. Farmers are deeply interested in good roads. Nearness to market, and facilities for all other kinds of communication, are worth a great deal, often materially affecting the price of land and its products. The difference between traveling ten miles through deep mud, at two miles per hour, with half a load, and traveling ten miles over a fine road, at five miles per hour, with a full load, should not be forgotten.

"In the absence of such facilities," says Gillespie, "the richest productions of nature waste on the spot of their growth. The luxuriant crops of our western prairies are sometimes left to decay on the ground, because there are no rapid and easy means of conveying them to market. The rich mines in the northern part of the State of New

York are comparatively valueless, because the roads among the mountains are so few and so bad, that the expense of the transportation of the metal would exceed its value. So, too, in Spain it has been known, after a succession of abundant harvests, that the wheat has actually been allowed to rot, because it would not repay the cost of carriage." Again, " When the Spanish government required a supply of grain to be transferred from Old Castile to Madrid, 30,000 horses and mules were necessary for the transportation of four hundred and eighty tons of wheat. Upon a broken-stone road of the best sort, *one-hundredth* of that number could easily have done the work." He further adds, in speaking of the improvements in roads made by Marshal Wade, in the Scottish Highlands, " His military road is said to have done more for the civilization of the Highlands than the preceding efforts of all the British monarchs. But the later roads, under the more scientific direction of Telford, produced a change in the state of the people which is probably unparalleled in the history of any country for the same space of time. Large crops of wheat now cover former wastes; farmers' houses and herds of cattle are now seen where was previously a desert; estates have increased seven-fold in value and annual returns; and the country has been advanced at least one hundred years."

THE WEDGE.

The wedge is a double inclined plane, the power being applied at the back to urge it forward. It becomes more and more powerful as it is made more acute; but, on account of the enormous amount of friction, its exact power can not be very accurately estimated. It is nearly always urged by successive blows of a heavy body, the momentum of which imparts to it great force.

All cutting and piercing instruments, as knives, scissors,

chisels, pins, needles, and awls, are wedges. The degree of acuteness must be varied according to circumstances; knives, for instance, which act merely by pressure, may be made with a much sharper angle than axes, which strike a severe blow. For cutting very hard substances, as iron, the edge must be formed with a still more obtuse angle.

The utility of the wedge depends on the friction of its surfaces. In driving an iron wedge into a frozen or icy stick of wood, as every chopper has observed, the want of sufficient friction causes it immediately to recoil, unless it be previously heated in the fire. The efficacy of nails depends entirely on the friction against their wedge-like faces.

THE SCREW

The screw may be regarded as nothing more than an inclined plane winding round the surface of a cylinder (fig. 74). This may be easily understood by cutting a piece of paper in such a form that its edge, *a b* (fig. 75), may represent the inclined plane; then, beginning at the wider end, and wrapping it about the cylindrical piece of wood, *c*, the upper edge of the paper will represent the thread of the screw.

Fig. 74

Although the friction attending the use of the screw is considerable, and without it it would not retain its place, yet the slope of its inclined thread being so gradual, it possesses great power. This power is multiplied to a still greater degree by the

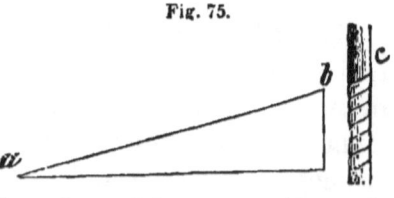

Fig. 75.

lever which is usually employed to drive it, *a* (fig. 76). If, for example, a screw be ten inches in circumference, and its thread half an inch apart, it exerts a force twenty times as great as the moving power. If it be moved by a lever twenty times as long as the diameter of the screw, here is another increase of twenty times in force. Multiplying 20 by 20 gives 400, the whole amount gained by this combination, and by which a man applying one hundred pounds in force could exert a pressure equal to twenty tons. About one-third or one-fourth of this should, however, be deducted for friction.

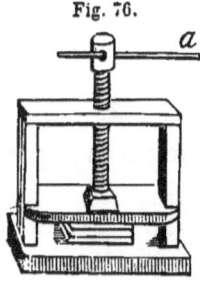

Fig. 76.

Screw and lever combined.

When the screw is combined with the wheel and axle (fig. 77), it is capable of exerting great power, which may be readily calculated by multiplying the power of the screw and its lever into the power of the wheel and axle.

THE KNEE-JOINT POWER.

The *knee-joint* or *toggle-joint* is usually regarded as a compound lever, and consists of two rods connected by a turning joint, as represented in fig. 78. The outer end of one of the levers is fixed to a solid beam, and the other connected with a movable block. When the joint a is forced in the direction indicated by the arrow, it produces a powerful pressure upon the movable block, which increases as the lever approaches a straight line. This is easily understood by the rule of virtual velocities, for the force moves with a velocity many times greater than the

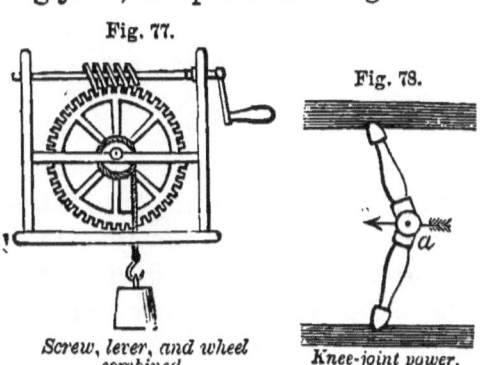

Fig. 77. *Screw, lever, and wheel combined.*

Fig. 78. *Knee-joint power.*

power given to the block, and this relative difference increases as the joint is made straighter.

This power is made use of in the lever printing-press, where the greatest force is given just as the pressure is completed. Another example occurs in the *Lever Washing-machine* (fig. 79), which is worked by the alternating motion of the handle, A, pressing a swinging board, per-

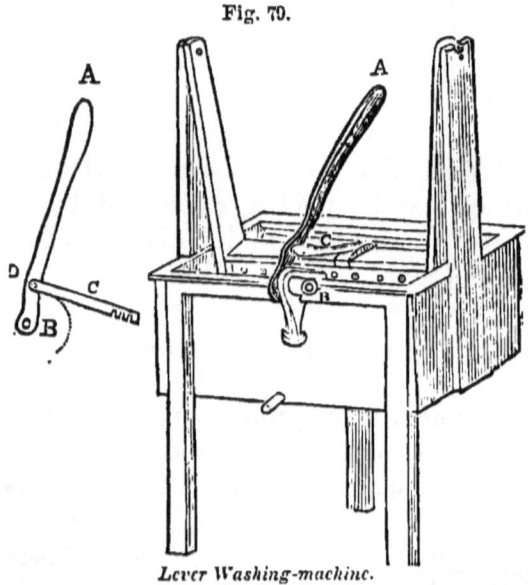

Fig. 79.

Lever Washing-machine.

forated with holes, with great force against the clothes next to one side of the water-box. Like the printing-press, this machine exerts the greatest power just as the motion of the lever is completed, and at the time it is most needed. The same principle is exhibited in *Kendall's Cheese-press* (fig. 80), where the lever and the wheel and axle are combined with the two knee-joints, one on each side of the press, drawing down a cross-beam upon the cheese with a greatly multiplied power.

Dick's Cheese-press (fig. 83), operates on a similar principle. Figs. 81–2 show the structure of its working

Fig. 80.

Kendall's Cheese-press.

part, the dotted lines indicating the position of the lever, which is inserted into a roller or axle, and, by turning, drives the movable iron blocks asunder, and raises the cheese against the broad screw-head above, as shown in fig. 82. In fig. 81, the raised lever shows that the blocks are at first near together, but are crowded asunder as the lever is pressed downward. This cheese-press is made of cast-iron, and has great power; to try it, weights were increased upon the lever, until the iron frame broke with a force equal to sixteen tons.

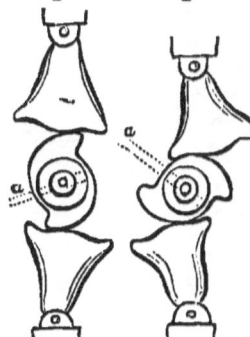

Fig. 82.   Fig. 81.

The power exerted by a *rolling-mill*, where bars of iron are flattened in their passage between two strong rollers, is precisely like that of the knee-joint. The only

Fig. 83.

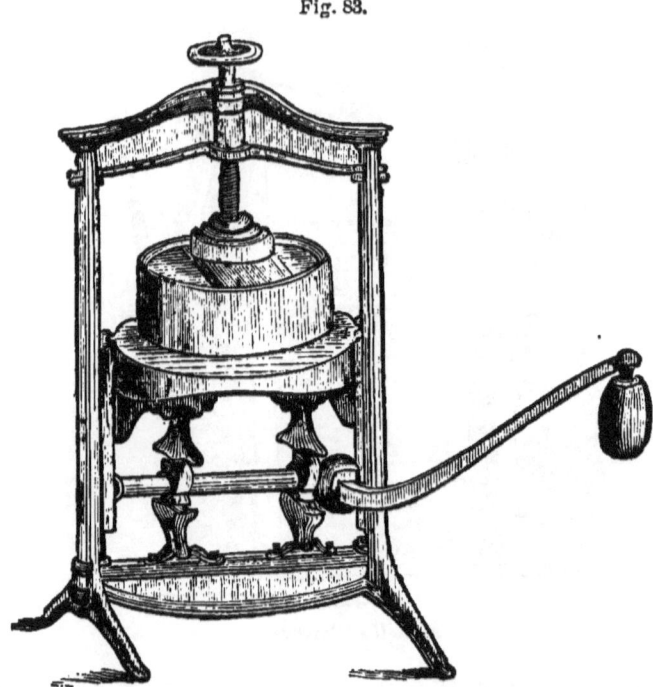

Dick's cast-iron Cheese-press.

difference is, that the rollers, which may be considered as a constant succession of levers coming into play as they revolve, are both fixed, and consequently the bar has to yield between them (fig. 84). The greatest power is exerted just as the bar receives the last pressure from the rollers. The most powerful and rapidly-working straw-cutters are those which draw the straw or hay between two rollers, one of which is furnished with knives set around it parallel with its axis, and cutting on the other, which is covered with untanned ox-hide (fig. 85).

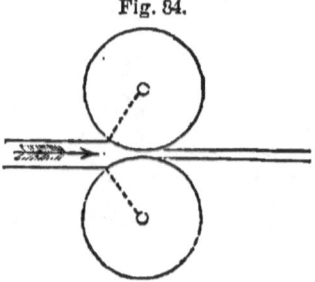

Fig. 84.

Principle of the knee-joint in the rolling-mill.

STRAW CUTTERS.

Fig. 85.

Hide Roller Straw Cutter.

CHAPTER V.

APPLICATION OF MECHANICAL PRINCIPLES IN THE STRUCTURE OF SIMPLE IMPLEMENTS AND PARTS OF MACHINES.

In contriving the more difficult and complex machines, the principles of mechanics must be closely studied, to give every part just that degree of strength required, and to render their operation as perfect as possible. But in making the more common and simple implements of the farmer, mere guess-work too often becomes the only guide. Yet it is highly useful to apply scientific knowledge even in the shaping of a hoe handle or a plow-beam.

The simplest tool, if constantly used, should be formed with a view to the best application of strength. The laborer who makes with a common hoe two thousand strokes an hour, should not wield a needless ounce. If any part is heavier than necessary, even to the amount of half an ounce only, he must repeatedly and continually lift this half ounce, so that the whole strength thus spent would be equal, in a day, to twelve hundred and fifty pounds, which ought to be exerted in stirring the soil and destroying weeds. Or, take another instance: A farm wagon usually weighs nearly half a ton; many might be

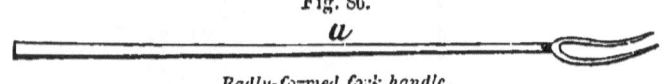

Fig. 86.

Badly-formed fork handle.

reduced fifty pounds in weight by proportioning every part exactly to the strength required. How much, then, should we gain here? Every farmer who drives a wagon with its needless fifty pounds, on an average of only five miles a day, draws an unnecessary weight every year equal to the conveyance of a heavy wagon-load to a distance of forty miles.

Now a knowledge of mechanical science will often enable the farmer, when he selects and buys his implements, to judge correctly whether every part is properly adapted to the required strength. We shall suppose, for instance, that he intends to purchase a common pitchfork. He finds them differently formed, although all are made of the

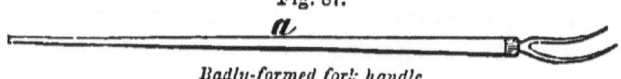

Fig. 87.

Badly-formed fork handle.

best materials. The handles of some are of equal size throughout. Some are smaller near the fork, as in fig. 86, and others are larger at the same place, as in fig. 87. Now, if he understands the principle of the lever, he knows that both of these are wrongly made, for the right hand placed at a is the fulcrum, where the greatest strength is needed, and therefore the one represented by fig. 88 is both stronger and lighter than the others.

Again, *hoe handles*, not needing much strength, chiefly require lightness and convenience for grasping. Hence, in selecting from two such as are represented in the annexed figures, the one should be chosen which is lightest near

Fig. 88.

Well-formed fork handle.

the blade, nearly all the motion being in that direction, because the upper end is the *centre* of motion. The right hand, at *a*, acting partly as the fulcrum, the hoe handle should be slightly enlarged at that place. Fig. 89 represents a well-formed handle; fig. 90, a clumsy one. Rake handles should be made largest at the middle, or where the right hand presses. Rake-heads should be much larger at the centre, and tapering to the ends, where the stress is least, the two parts operating as two distinct lev-

Fig. 89.

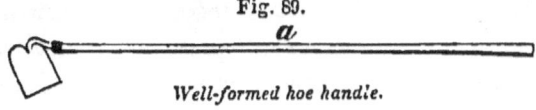

Well-formed hoe handle.

ers, acting from the middle. Wood *horse-rakes* might be made considerably lighter than they usually are by observing the same principles. The greatest strength required for *plow-beams* is at the junction with the mould-board, and the least near the forward end, or furthest from the fulcrum or centre of motion.

Now it may be that the farmer who has had much experience may be able to judge of all these things without

Fig. 90.

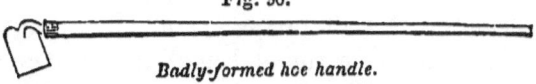

Badly-formed hoe handle.

a knowledge of the science. But this scientific knowledge would serve to strengthen his experience, and enable him to judge more accurately and understandingly by showing him the reasons; and in many cases, where *new* implements were introduced, he might be enabled to form a good judgment before he had incurred all the expense and losses of unsuccessful trials.

Even so simple a form as that of an ox-yoke is often made unnecessarily heavy. Fig. 91 represents one that is faulty in this respect, by having been cut from a piece of

Fig. 91.

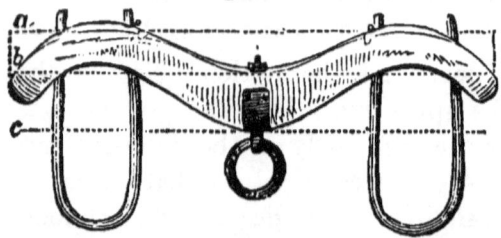

timber as wide as the dotted lines $a\ c$; and being thus weakened, it requires to be correspondingly large. Fig. 92 is equally strong, much lighter, and is easily made from a stick of timber only as wide as $a\ b$ in the former figure.

In the heavier machines, it is necessary to know the degree of taper in the different parts with accuracy. A thorough knowledge of science is needed to calculate this

Fig. 92.

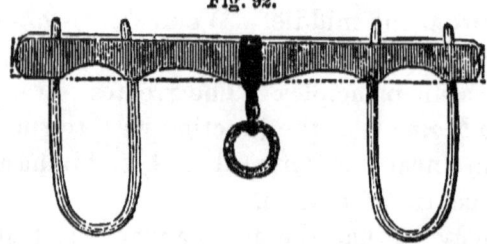

with precision, but a superficial idea may be given by cuts. If a bar of wood, formed as in a (fig. 93), be fixed in a wall of masonry, it will possess as much strength to support a weight hung on the end as if it were the same size throughout, as b. The first is equally strong with the second, and much lighter.* The same form doubled must

---

* The simple style of this work precludes an explanation of the mode of calculation for determining the exact form. Where the stick tapers only on one side, it is a common parabola; if on all sides, a cubic parabola.

be given if the bar is supported at the middle, with a weight at each end, or with the weight at the middle, supported at each end, as *c*. This form, therefore, is a proper one for many parts of implements, as the bars of whiffle-trees, the rounds of ladders, string-pieces of bridges, and any cross-beams for supporting weights. The proper form for rake-teeth and fence-posts, the pressure being nearly alike on all parts, is nearly that of a long wedge, or with a straight and uniform taper. Therefore a fence-post of equal size throughout contains nearly twice as much timber as is needed for strength only.

The form of these parts must, however, be modified to suit circumstances; as whiffle-trees must be large enough

Fig. 93.

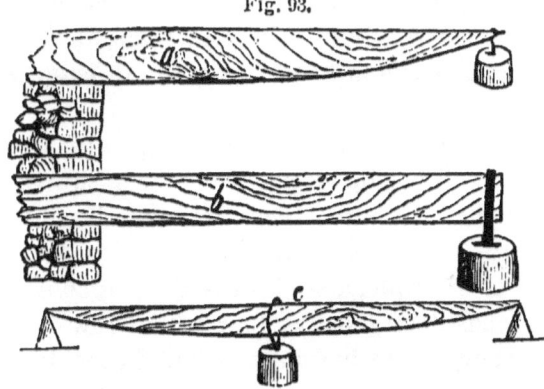

at the ends to receive the iron hooks, wagon-tongues for ironing at the end, and spade handles for the easy grasp of the hand.

The axle-trees of wagons must be made not only strong in the middle, or at centre of pressure, but also at the entrance of the hub; because the wheels, when thrown sidewise in a rut, or on a sideling road, operate as levers at that point. *a* and *b* (fig. 94), show the manner in which the axles of carts may be rendered lighter without lessening the strength, *a* being the common form, and *b* the improved one.

Sometimes several forces act at once on different parts. For example, the spokes of wagon-wheels require strength at the hub for stiffening the wheel; they must be strong in the middle to prevent bending, and large enough at

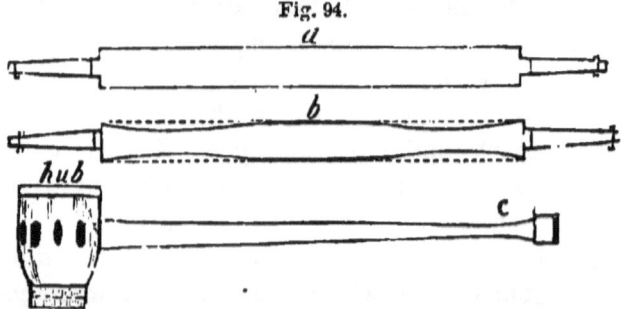

Fig. 94.

the outer ends, where they are soonest weakened by decay. Hence there should be nearly a uniform taper, slightly larger at the middle, and with an enlargement at the outer end, as *c* (fig. 94).

A very useful rule in practice, in giving strength to structures, is this: The strength of every square beam or stick to support a weight increases exactly as the width increases, and also exactly as the *square of the depth* increases. For example, a stick of timber eight inches wide and four inches deep (that is, four inches thick), is exactly twice as strong as another only four inches wide, and with the same depth. It is twice as wide, and consequently twice as strong; that is, its strength increases just as the width increases, according to the rule given. But where one stick of timber is twice as *deep*, the width being the same, it is *four times* stronger; if three times as deep, it is *nine times* stronger, and so on. Its strength increases as the *square* of the depth, as already stated. The same rule will show that a board an inch thick and twelve inches wide will be twelve times as strong when edgewise as when lying flat. Hence the increase in strength given to whiffle-trees, fence-posts, joists, rafters, and string-pieces to farm-bridges, by making them narrow and deep.

Again, the strength of a *round* stick increases as the cube of the diameter increases; that is, a round piece of wood three inches in diameter is eight times as strong as one an inch and a half in diameter, and twenty-seven times as strong as one an inch in diameter. This rule shows that a fork handle an inch and a half in diameter at the middle is as much stronger than one an inch and a quarter in diameter, as seven is greater than four. Now this rule would enable the farmer to ascertain this without breaking half a dozen fork handles in trying the experiment, and it would enable the manufacturer to know, without

Fig. 95.

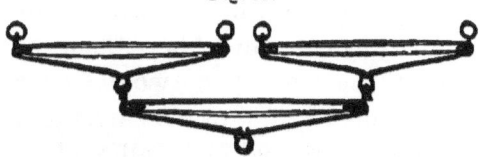

the labor of trying many experiments, that if he makes a fork handle an inch and a half at the middle, tapering a quarter of an inch toward the ends, it will enable the workman to lift with it nearly twice as much hay as with one an inch and a quarter only through its whole length.

A mode of adding strength to light bars of wood, by means of braces, is shown in fig. 95, representing light whiffle-trees, stiffened by iron rods in a simple manner. The same method is sometimes adopted to advantage in making light fruit ladders, and for other purposes.

CHAPTER VI.

FRICTION.

The subject of friction has been postponed, or merely alluded to, to prevent the confusion of considering too many things at once. As it has an important influence on the action of machines, it is worthy of careful investigation.

4*

It is familiar to most persons, that when two surfaces slide over each other while pressing together, the minute unevenness or roughness of their surfaces causes some obstruction, and more or less force is required. This resistance is known as *friction*.

ROLLING FRICTION.

The term is also applied to the resistance of one body *rolling over* another. This may be observed in various degrees by rolling an ivory ball successively over a carpet, a smooth floor, and a sheet of ice; the same force which would impel it only a few feet on the carpet would cause it to move as many yards on a bare floor, and a still greater distance on the ice. The two extremes may be seen by the force required to draw a carriage on a deep sandy or loose-gravel road, and on a rail-road.

NATURE OF FRICTION.

If two stiff bristle brushes be pressed with their faces together, they become mutually interlocked, so that it will be quite difficult to give them a sliding motion. This may be considered as an extreme case of friction, and serves to show its nature. In two pieces of coarse, rough sandstone, or of roughly-sawed wood, asperities interlock in the same way, but less in degree; a diminished force is consequently required in moving the two surfaces against each other. On smoothly planed wood the friction is still less; and on polished glass, where the unevenness can not be detected without the aid of a powerful magnifying glass, it is reduced still further in degree.

ESTIMATING THE AMOUNT OF FRICTION.

In order to determine the exact amount of friction between different substances, the following simple and ingenious contrivance is adopted: An inclined plane, *a b* (fig. 96), is so formed that it may be raised to any desired height by means of the arc of a circle and a screw. Lay a flat surface of the substance we wish to examine upon this inclined plane, and another smaller piece or block of the same substance upon this surface; then raise the plane until it becomes just steep enough for the block to slide down by its weight. Now, by measuring the degree of slope, we know at once the amount of friction. Suppose, for example, the two surfaces be smoothly-planed wood: it will be found that the plane must be elevated about half as high as its length; therefore we know, by the

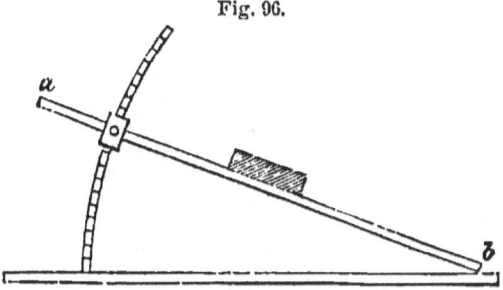

Fig. 96.

properties of the inclined plane, heretofore explained, that it requires a force equal to one-half the weight of the wooden block to slide it over a smooth wooden surface. Some kinds of wood have more friction than others, but this is about the average.*

From the result of this experiment we may learn that to slide any object of wood across a floor requires an amount of strength equal to one-half the weight of the object. A heavy box, for instance, weighing two hundred pounds, can not be moved without a force equal to one hundred pounds. It also shows the impropriety of placing

---

* These experiments may be made with tolerable accuracy, by hooking a spring-balance into any object of known weight, and then observing the comparative force as measured by the balance, to draw it over a perfectly level surface.

a heavy load upon a sled in winter for crossing a bare wooden bridge or a dry barn floor, the friction between cast-iron sleigh-shoes and rough sanded plank being nearly equal to one-third of the whole weight.* Hence a load of one ton (including the sled) would require a draught equal to more than six hundred pounds, which is too much for an ordinary single team. On bare unfrozen ground the friction would be still greater. On a plank bridge, with runners wholly of wood, it would be equal to half the load. All these facts may be readily proved by actually placing the sled on slopes of plank and of earth, and by observing the degree of steepness required for sliding down by its own weight.

In a similar way, we are enabled easily to ascertain the force required to draw a wagon upon any kind of level surface. Suppose, for example, that we wish to determine the precise amount of force for a wagon weighing, with its load, one ton, on a plank road. Select some slight descent, where the wagon will barely run with its own weight. Ascertain by a *level* just what the degree of descent is; then divide the weight of the wagon by the degree of the slope, and we shall have the force sought for. To make this rule plainer by an example: It will be found that a good, newly-laid plank track, if it possess a descent of only one foot in fifty feet distance, will be sufficient to give motion to an easy-running wagon; therefore we know that the strength required to draw it on a level will be only one-fiftieth part of a ton, or forty pounds.

The resistance offered to the motion of a wagon by a Macadam road, by a common dry road, and by one with six inches of mud, may be readily determined in the same way by selecting proper slopes for the experiment. If by such trials as these the farmer ascertains the fact that a

---

* On clean hard wood, with polished metallic shoes, the friction would be much less, or a fourth or fifth.

few inches of mud are sufficient to retard his wagon so much that it will not run of its own weight down a slope of one foot in four (and few common roads are ever steeper), then he may know that a force equal to one-fourth the whole weight of his wagon and load will be required to draw it on a level over a similar road—that is, the enormous force of five hundred pounds will be needed for one ton, of which many wagons will constitute nearly one-half. Hence he can not fail to see the great importance, for the sake of economy, and humanity to his team, of providing roads, whether public or private, of the hardest and best materials.

RESULTS WITH THE DYNAMOMETER.

Another mode of determining the resistance of roads is by means of the *Dynamometer.** It resembles a *spring-balance*, and one end is fastened to the wagon and the other end connected with the horses. The force applied is measured on a graduated scale, in the same way that the weight of any substance is measured with the spring-balance. A more particular description of this instrument will be given hereafter.

Careful experiments have been made with the dynamometer to ascertain accurately the resistance of various kinds of roads. The following are some of the results:

On a new gravel road, a horse will draw eight times as much as the force applied; that is, if he exerts a force equal to one hundred and twenty-five pounds, he will draw half a ton on such a road, including the weight of the wagon, *the road being perfectly level.*

On a common road of sand and gravel, sixteen times as much, or one ton.

On the best hard-earth road, twenty-five times as much, or one and a half tons.

---

* From two Greek words, *dunamis*, power, and *metreo*, to measure.

On a common broken-stone road, twenty-five to thirty-six times as much, or one and a half to two and a quarter tons.

On the best broken-stone road, fifty to sixty-seven times as much, or three to four tons.

On a common plank-road, clean, fifty times as much, or three tons.

On a common plank-road, covered thinly with sand or earth, thirty to thirty-five times as much, or about two tons.

On the smoothest oak plank-road, seventy to one hundred times as much, or four and a half to six tons.

On a highly-finished stone track-way, one hundred and seventy times as much, or ten and a half tons.

On the best rail-road, two hundred and eighty times as much, or seventeen and a half tons.

The firmness of surface given to a broken-stone road by a paved foundation was found to lessen the resistance about one-third.

On a broken-stone road it was found that a horse could draw only about two-thirds as much when it was moist or dusty as when it was dry and smooth; and when muddy, not one-half as much. When the mud was thick, only about one quarter as much.

The character of the vehicle has an influence on the draught. Thus, a cart, a part of the load of which is supported by the horse, usually requires only about two-thirds the force of horizontal draught needed for wagons and carriages. On rough roads the resistance is slightly diminished by springs.

On soft roads, as earth, sand, or gravel, the number of pounds draught is but little affected by the speed; that is, the resistance is no greater in driving on a trot than on a walk; but on hard roads it becomes greater as the velocity increases. Thus a carriage on a dry pavement requires one-half greater force when the horses are on a trot than on a walk; but on a muddy road the difference between the two rates of speed is only about one-sixth. On a railroad, where a draught of ten pounds will draw a ton ten miles an hour, the resistance increases so much at a high degree of speed as to require a force of fifty pounds per ton at sixty miles an hour—that is, it would require five times as much actual power to draw a train one hundred miles at the latter rate as at the former; but as the speed is six times as great, the actual force *during a given time* would be five times six, or thirty times as great.

WIDTH OF WHEELS.

Wheels with wide tire run more easily than narrow tire, on soft roads; on hard, smooth roads, there is no sensible difference. Wide tire is most advantageous on gravel and new broken-stone roads, both by causing the vehicles to run more easily, and by improving the surface. For the latter reason, the New York turnpike law allows six-inch wheels to pass at half price, and twelve-inch wheels to pass free of toll. Wheels with broad tire on a farm would pass over clods, and not sink between them; or would only press the surface of new meadows, without cutting the turf. But where the ground becomes muddy, the mud closes on both sides of the rim, and loads the wheels. On clayey soils, narrow tire unfits the roads for broad wheels. For these reasons, broad wheels are decidedly objectionable for clayey or soft soils, and they are chiefly to be recommended for broken-stone roads, and gravelly, or dry, sandy localities. They are also much better for the wheels of sowing or drilling machines, which only pass over mellowed surfaces.

The larger the wheels are made, the more easily they run; thus a wheel six feet in diameter meets with only half the resistance of a wheel three feet in diameter.

A flat piece of wood, sliding on one of its broad surfaces, is subject to the same amount of friction as when sliding upon its edge. Hence the friction is the same, provided the pressure be the same, whether the surface be small or large.* Or, in other words, if the surfaces are the same, a double pressure produces a double amount of friction; a triple pressure, a triple amount, and so on.

A narrow *sleigh-shoe* usually runs with least force, for two reasons: first, its forward part cuts with less resistance through the snow; and, secondly, less force is required to pack the narrow track of snow beneath it. The only instance in which a wide sleigh-shoe would be best, is where a crust exists that would bear it up, and through which a narrow one would cut and sink down.

VELOCITY.

Friction is entirely independent of velocity; that is, if a force of ten pounds is required to turn a carriage wheel, this force will be ten pounds, whether the carriage is driven one or five miles per hour. Of course, it will require five times as much force to draw five miles per hour, because five times the distance is gone over; but, measured by a dynamometer or spring-balance, the pressure would be the same. In precisely the same way, the weight of a stone remains the same, whether lifted slowly or quickly. If the friction of the wheels of a wagon on their axles be equal to ten pounds, driving the horse fast or slowly will not increase or diminish it. But fast driving will require more strength, for the same reason that a man would need more strength to carry a bag of wheat up two flights of stairs than one, in one minute of time.

FRICTION AT THE AXLE.

A carriage wheel, or any other wheel revolving on an

---

* Generally speaking, this is very nearly correct; but when the pressure is intense, the friction is slightly less on the smaller surface.

axle, will run more easily as the axle is made smaller. This is not owing to the rubbing surfaces being less in size, as some mistakenly suppose, for it has just been shown that this makes very little or no difference, provided the pressure is the same; but it is owing to the leverage of the wheel on the friction at the axis; and the smaller the axle, the greater is this leverage; for, if the axle, a (fig. 97), be six inches in circumference, and the wheel, $b\ c$, be ten feet in circumference, then the outer part of the wheel will move twenty times further than the part next the axle. Therefore, according to the rule of virtual velocities (already explained,) one ounce of force at the rim of the wheel will overcome twenty ounces of friction at the axle; or if the axle were twice as large, then, according to the same rule, it would require two ounces to overcome the same friction acting between larger surfaces.

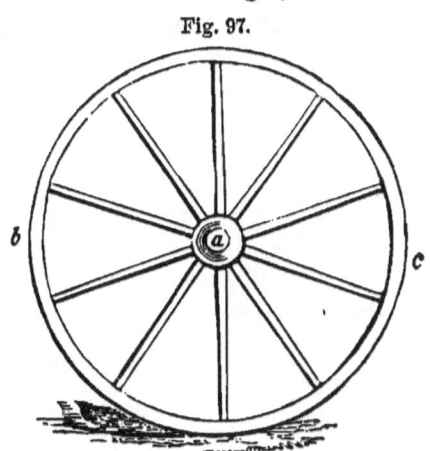

Fig. 97.

For this reason, large wheels in wheel-work for multiplying motion, if not made too heavy, run with less force than smaller ones, the power acting upon a larger lever. Horse-powers for thrashing-machines, consisting chiefly of a large, light crown-wheel, well stiffened by brace-work, have been found to run with remarkable ease; a good example of which exists in what is known as *Talpin's* horse-power, when made in the best manner.

FRICTION-WHEELS.

On the preceding principle, *friction-wheels* or *friction-rollers* are constructed, for lessening as much as possible the friction of axles in certain cases. By this contrivance, the axle, *a* (fig. 98), instead of revolving in a simple hole or cavity, rests on or between the edges of two other wheels. As the axle revolves, the edges turn with it, and the rubbing of surfaces is only at the axles of these two wheels. If, therefore, these axles be twenty times smaller than the wheels, the friction will be only one-twentieth the amount without them. This contrivance has been strongly recommended and considerably used for the cranks of grindstones (fig. 99), but it was not found to answer the intended purpose so well as was expected, for the very plain reason that, in using a grindstone, nearly all the friction is at the circumference, or between the stone and the tool, which friction-wheels could not, of course, remove.

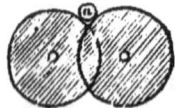

Fig. 98.

Friction-wheels.

Fig. 99.

Grindstone on Friction-wheels.

LUBRICATING SUBSTANCES.

Lubricating substances, as oil, lard, and tallow, applied to rubbing surfaces, greatly lessen the amount of friction, partly by filling the minute cavities, and partly by separating the surfaces. In ordinary cases, or where the machinery is simple, those substances are best for this purpose which keep their places best. Finely-powdered black-lead, mixed with lard, is for this reason better for greasing carriage wheels than some other applications. Drying oils, as linseed, soon become stiff by drying, and are of little service. Olive oil, on the contrary, and some animal oils, which scarcely dry at all, are generally preferred. To obtain the full benefit of oil, the application must be frequent.

According to the experiments made with great care by Morin, at Paris, the friction of wooden surfaces on wooden surfaces is from one quarter to one-half the force applied; and the friction of metals on metals, one-fifth to one-seventh—varying in both cases with the kinds used. Wood on wood was diminished by lard to about one-fifth to one-seventh of what it was before; and the friction of metal on metal was diminished to about half what it was before; that is, the friction became about the same in both cases after the lard was applied.

To lessen the friction of wooden surfaces, lard is better than tallow by about one-eighth or one-seventh; and tallow is better than dry soap about as two is to one. For iron on wood, tallow is better than dry soap about as five is to two. For cast-iron on cast-iron, polished, the friction with the different lubricating substances is as follows:

Water	31
Soap	20
Tallow	10
Lard	7
Olive oil	6
Lard and black-lead	5

When bronze rubs on wrought iron, the friction with lard and black-lead is rather more than with tallow, and about one-fifth more than with olive oil. With steel on bronze, the friction with tallow and with olive oil is about one-seventh less than with lard and black-lead.

As a general rule, there is least friction with lard when hard wood rubs on hard wood; with oil, when metal rubs on wood, or metal on metal—being about the same in each of all these instances.

In simple cases, as with carts and wagons, where the

friction at the axle is but a small portion of the resistance,* a slight variation in the effects in the lubricating substance is of less importance than retaining its place. In more complex machinery, as horse-powers for thrashing-machines, friction becomes a large item, unless the parts are kept well lubricated with the best materials.

Leather and hemp bands, when used on drums for wheel-work, should possess as much friction as possible, to prevent slipping, thus avoiding the necessity of tightening them so much as to increase the friction of the axles. Wood with a rough surface has one-half more friction than when worn smooth; hence moistening and rasping small drums may be useful. Facing with buff leather or with coarse thick cloth also accomplishes a useful purpose. It often happens that wetting or oiling bands will prevent slipping, by keeping their surfaces soft, and causing them to fit more closely the rough surface of the drum.

ADVANTAGES OF FRICTION.

Although friction is often a serious inconvenience, or loss, in lessening the force of machines, there are many instances in which it performs important offices in nature and in works of art. "Were there no friction, all bodies on the surface of the earth would be clashing against each other; rivers would dash with an unbounded velocity, and we should see little besides collision and motion. At present, whenever a body acquires a great velocity, it soon loses it by friction against the surface of the earth. The friction of water against the surfaces it runs over soon reduces the rapid torrent to a gentle stream; the fury of

---

* If the friction at the axle be one-twelfth of the force, and the diameter of the wheels ten times as great as the diameter of the axle, the friction at the axles will be reduced to one-twelfth of a tenth, or one hundred and twentieth part of the force, according to the law of virtual velocities as applied to the wheel and axle.

the tempest is lessened by the friction of the air on the face of the earth, and the violence of the ocean is subdued by the attrition of its own waters.

"Its offices in the works of art are equally important. Our garments owe their strength to friction, and the strength of ropes depends on the same cause; for they are made of short fibres pressed together by twisting, causing a sufficient degree of friction to prevent the sliding of the fibres. Without friction, the short fibres of cotton could never have been made into such an infinite variety of forms as they have received from the hands of ingenious workmen."* Deprived of this retaining force, the parts of stone walls, piles of wood and lumber, and the loads of carts and wagons, as well as the wheels themselves, would slide without restraint, as if their surfaces were of the most icy smoothness, and walking without support would be impossible.

The tractive power of locomotives depends on the friction between the wheels and iron rails, which is equal to about one-fifth of the weight of the engine; that is, a locomotive weighing twenty-five tons will draw with a force of five tons, without producing slipping of the wheels.

CHAPTER VII.

PRINCIPLES OF DRAUGHT.

An examination of the nature or laws of friction enables us to ascertain the best line of draught for teams when attached to wagons and carriages. If there were no friction whatever upon the road, the best direction for the

---

* Encyclopædia Americana.

traces would be parallel with its surface, that is, on a level; but as there is always some friction, the line of draught should be a little rising, so as to tend to lessen the pressure of the wheels on the road.

Now this upward direction of the draught *should always be exactly of such a slope, that if the same slope were given to the road, the wagon would just descend by its weight.* The more rough or muddy the road is, the steeper should be this line of draught or direction of the traces.* On a good common road it would be much less, and on a plank-road but slightly varied from a horizontal direction. On a rail-road the line should be about level. On good sleighing, some of the strength of the team is commonly lost by too steep a line of draught.

The reason of this rule may be understood by the following explanation: Let the obstruction, *a*, in the annexed figure (fig. 100) represent the friction the wheel constantly meets with in rolling over a common road. To overcome this friction, the wheel must rise in the direction of the dotted line. Therefore, if the force is made to pull in this direction, it will act more advantageously than in any other, because this is the course in which the centre of the wheel must move. Now if a downward slope were given to the road at this obstruction, the wheel and the obstruction would both be brought on a level, and the wheel would move with the slightest degree of force.

It will be understood from the preceding rule that a sled running on bare ground should be drawn by traces bearing upward in a large degree. The same remark will apply to the plow, which slides upon the ground in a similar way, with the pressure of the turning sod as a load. Hence

---

* Provided the wheels are not made smaller for this purpose, increasing their resistance.

the reason that a great saving of strength results from the use of short traces in plowing. An experiment was tried for the purpose of testing this reasoning; first, with traces of such length that the horses' shoulders were about ten feet from the point of the plow; and secondly, with the distance increased to about fifteen feet. With the short traces a strength was required equal to $2\frac{1}{4}$ cwt., but with the long traces it amounted to $3\frac{1}{2}$ cwt.

But the draught-traces may be made too short. When this is the case, the plow is necessarily thrown too much upon its point, to keep it from flying out of the ground, by which means it works badly in turning the furrow. In addition to this evil, the plowman is compelled to bear down heavily, adding to the friction of the sole on the bottom of the furrow, and greatly increasing his labor.

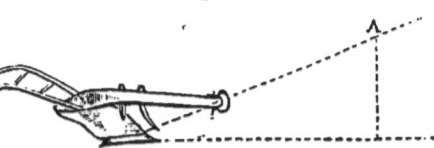

Fig. 101.

The line of draught should be so adjusted that the plow may press equally all along on its sole or bottom, which will cause it to run evenly and with a steady motion.

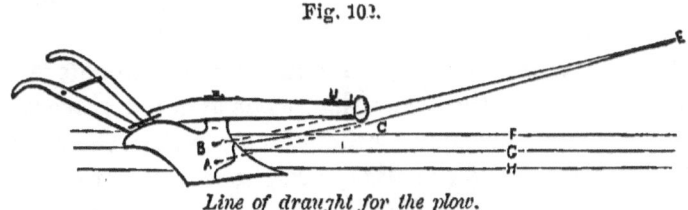

Fig. 102.

Line of draught for the plow.

This end will be effected by giving the traces or draught-chain just such a length that the share of the plow (or centre of resistance), the clevis, and the point of draught at the horses' shoulders (or the ring of the ox-yoke) shall all form a straight line. This is shown in the annexed figure, where A is the place of the ox-ring or of the forward extremity of the traces (fig. 101).

The centre of resistance will vary with the depth of

plowing. When the furrow is shallow (as shown by the lines G H, fig. 102), the centre of resistance will be at A, requiring the team to be fastened to the lower side of the clevis, C; but when the depth is greater (as shown by F H), the centre of resistance will be at B, requiring a higher attachment to the clevis; the point of draught, E, remaining the same in both cases.

So great is the difference between an awkward and skillful adjustment of the draught to the plow, that some workmen with a poor implement have succeeded better than others with the best; and plows of second quality have sometimes, for this reason, been preferred to those of the most perfect construction.

COMBINED DRAUGHT OF ANIMALS.

When several animals are combined together, it is of great importance that they should be exactly matched in gait. Much force is often wasted when they draw unsteadily or unevenly. It is more difficult to divide the draught equally among several animals when placed one before the other, than when arrayed abreast, for some may hang back, and others do more than their share, unless a skillful driver is always on the watch. It also happens, when thus arranged, that the forward horses draw horizontally, while the hindmost one draws in a sloping line, and the line of draught between them thus being crooked, more or less force is lost. This may be, however, remedied in part by placing the taller animals forward, and the smaller behind.

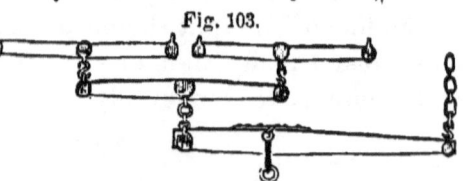

Fig. 103.

For these reasons, when only three horses are used, they should always be placed abreast. The force required for each may be rendered exactly equal by the whiffle-trees usually employed for this purpose, and represented in fig. 103, where two horses are attached to the shorter end, and

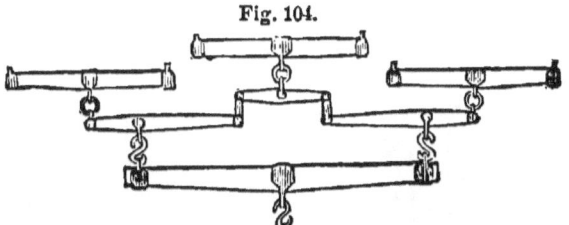

Fig. 104.

Whipple-tree for three horses.

the third to the longer end of the common bar. Another ingenious but more complex arrangement is shown by fig. 104, where also the central horse has only half the two others, by being attached to the longer ends of the intermediate bars. Another, and a more perfect contrivance, is *Potter's Three-horse Clevis*, re-

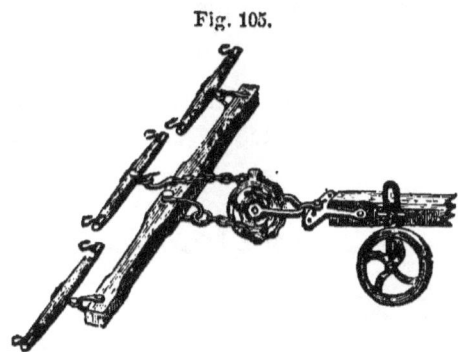

Fig. 105.

presented by fig. 105. It consists of two wheels together, one twice the diameter of the other, and each having a groove in which a chain runs. The chains are fastened to the respective wheels, so that the single horse draws on the larger wheel, against the two horses on the smaller. With common whiffle-trees, the relative draught of each horse is maintained only when they draw evenly; with Potter's there is no variation at any time. It is made by E. M. Potter, Kalamazoo, Mich. Fig. 106 represents the mode of attaching four horses in draught, their force being equalized by passing the chain round the wheel in the pulley-block, *a*, security being provided that the hindmost pair shall not encroach on the

5

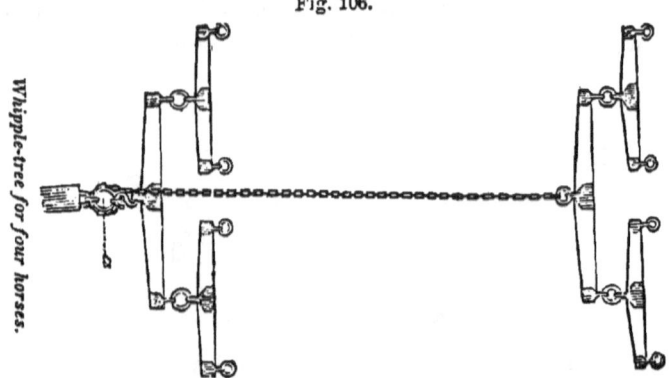

Fig. 106.

forward pair, by connecting the end of the chain at the same time to the plow.

WIER'S SINGLE-TREE.

This is used exclusively for plowing in orchards, and is worthy of notice here. The leather traces are hooked at

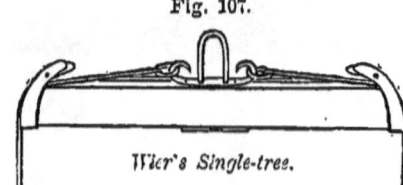

Fig. 107.
Wier's Single-tree.

the rear of the wooden bar, and, passing around the ends, prevent the possibility of being caught in the bark of the trees. The teamster may therefore drive as closely as he chooses without danger of injury. For this reason he is able to turn over the whole surface without leaving an unplowed strip along the row.

CONSTRUCTION AND USE OF THE DYNAMOMETER.

The dynamometer, or force-measurer, has been already briefly alluded to, but a more particular description will be useful. In the construction and selection of all machines and implements that require much power in their use, the dynamometer is indispensable, although at present but little known. As an example of its utility, the farmer may wish to choose between two plows which, so far as he can perceive, may do their work equally well; but this instrument, when applied, may show that the team must draw with a force equal to 400 pounds in moving one of them through the soil, while 300 pounds would be sufficient for the other. He would, therefore, select the one of easiest draught, and by doing so would save the labor of one day in four to his team, or twenty-five days in a hundred, which would be worth many times the cost of the trial. The same advantage might be derived in the selection of harrows, cultivators, horse-rakes, straw-cutters, and all other implements drawn by horses or worked by men. Again, the farmer may be in doubt in choosing between two thrashing-machines, which in other respects may work equally fast and well; but the dynamometer may show that one requires a severer exertion from the team, and consequently is less valuable for use.

The operation of this instrument may be readily under-

Fig. 108.

Dynamometer, or Force-measurer.

stood by fig. 108, where *b* represents the dynamometer,

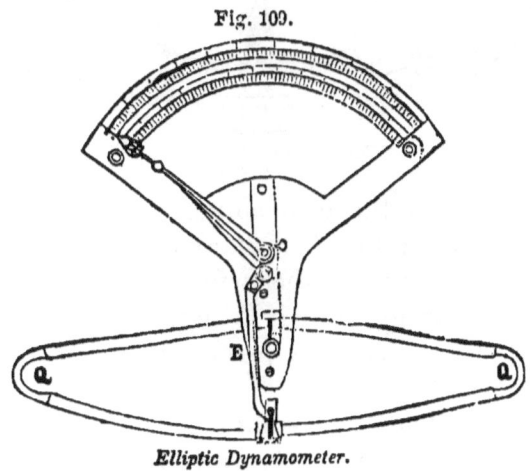

Fig. 109.

Elliptic Dynamometer.

made precisely similar to a large and stiff spring balance,

with one hook attached to the plow and the other to the whiffle-tree. The amount of force required to draw the plow is accurately measured on the scale by the index or pointer, *a*.

Sometimes the motion of this index is multiplied, or made greater and more easily seen, by means of a cog-wheel and rack-work; but this renders the instrument, at the same time, more complex.

Another form of this instrument is shown in fig. 109,

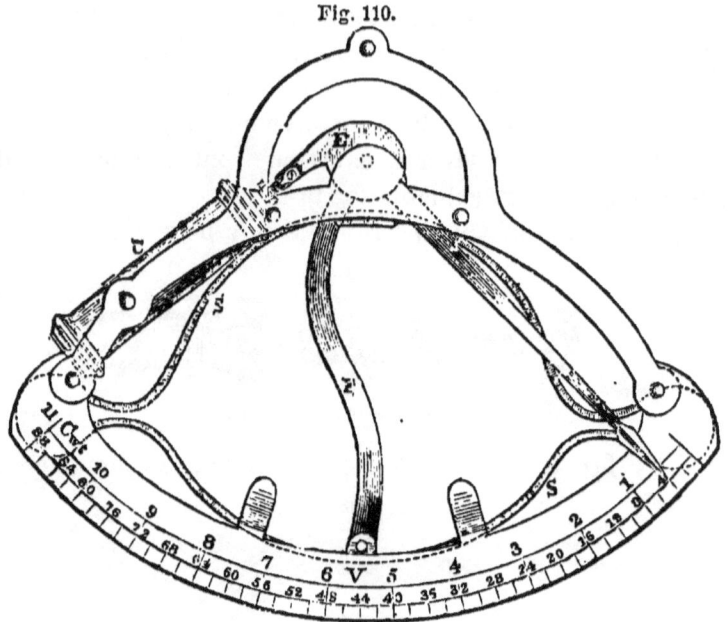

Fig. 110.

Elliptic Dynamometer, in compact form: S S, *spring;* F, *cross-lever for moving index.*

where the ends of the oval spring, Q Q, are attached to the plow and draught. The harder the force exerted by the team, the closer together will the sides of this spring be brought, causing the rod, E, to press against the index or pointer, and showing the precise degree of force on the circular scale.

An improvement, by rendering the instrument more compact, is shown in fig. 110, where S S is the spring, and directly over it is the graduated scale.

SELF-RECORDING DYNAMOMETER.

An inconvenience occurs in the use of the instruments now described from the rapid vibration of the index, resulting from the quick changes in the force, partly from inequalities in the soil, and partly from the unsteady motion of the horses. The vibration is sometimes so great that the index can hardly be seen, rendering it difficult to measure the average force. This inconvenience has been removed, in a great degree, by attaching to one end of the index, E (fig. 110), a piston working in a cylinder filled with oil, C; this piston has a small hole through it, through which the oil passes from one side to the other as the draught varies, but not fast enough to allow any sudden motion.

SELF-RECORDING DYNAMOMETER.

A less simple but more perfect instrument is the *Self-recording Dynamometer*, which marks accurately all the

Fig. 111.

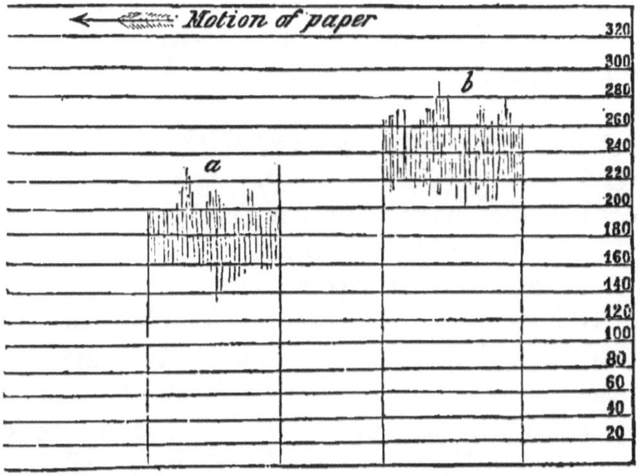

The markings of the Self-recording Dynamometer.

vibrations on a slip of paper while the plow is in operation. A pencil is fixed to the index, and presses, by means of a spring, against the paper, thus giving a true register of the force exerted. To prevent the pencil from constantly marking on the same line, the paper is made to move slowly in a side direction, so that all the vibrations are shown, as represented in fig. 111, and they may be accurately examined and read off at leisure, a and b representing the forces of two different plows, drawn through a single furrow across the field. The motion of the paper is effected by being placed on two rollers, one of which unwinds it from the other. This roller is made to turn by means of a wheel running on the ground, which gives motion to the roller through an endless chain, working a cog-wheel by means of an endless screw. The cylindrical dynamometer, shown in fig. 112, is used for this purpose, lengthwise upon which the two rollers are placed for holding the paper. With this instrument a permanent register might be made of the force required for different plows, with an accuracy not liable to dispute.

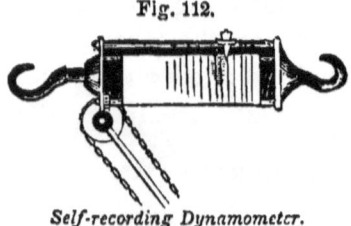

Fig. 112.

Self-recording Dynamometer.

WATERMAN'S DYNAMOMETER.

All difficulties have been completely overcome by the recent invention of H. Waterman, of Hudson, N. Y. His dynamometer was used with entire success at the Auburn reaper trial in 1866, and at the trial of plows at Utica, in 1867, under the Committee of the N. Y. State Agricultural Society. A full description of all the parts would require too much space for the character of this work; the following is a brief explanation of the mode of its operation:

This dynamometer is furnished with a spiral spring, like those we have already described, working a piston in a cylinder of water. To this, two dial plates are added one of which shows, by a slowly revolving index, the exact distance which the horses have traveled, without looking at in for a distance of more than five miles. The other dial plate gives a perfectly accurate record of the whole force expended from the commencement of the experiment to its termination. In other words, it takes all the different and varying forces, and adds them accurately in one aggregate or whole, seen at a glance on the dial plate under the eye.

We shall attempt a brief description of the modes by which the indexes on these two dial plates are moved.

The mode by which the distance traveled is recorded will be easily understood. A wheel one yard in circumference runs on the ground and communicates its motion by a cord, to a wheel attached to the dynamometer. This, by means of an endless screw and cog-work, moves the index slowly around the face, and thus records the distance traveled. There are two parts of this portion of the apparatus, which deserve a description. One is the wheel around which the cord passes in connection with the wheel which runs on the ground. It is very important that the exact number of the revolutions of this wheel should be maintained, as compared with those of the ground wheel. This is regulated as follows: The groove in this wheel is made by screwing together two beveled edged wheels, as shown in the annexed section, fig. 113. By placing thin paper between these two wheels, the width of the groove may be varied with the utmost accuracy, and the cord consequently let further in towards the centre. The other part which we desire to notice, although not original in this dynamometer, is the manner in which the index is carried around the face of the dial plate. There are two cog-wheels on the same axis, one with a hundred cogs, and the other with ninety-nine—both fitting into the same pinion. Consequently, when one has made the entire revolution, the other has fallen one cog behind, and a hundred revolutions are required for the index, placed upon one of them, to come around again so as to coincide with its first position.

The endless screw attached to the band-wheel already noticed moves one cog at every yard advanced, and the index passing around in a hundred revolutions, it is obvious that it will show 10,000 yards, or more than five miles.

We shall now attempt to describe that part of the machine which furnishes an accurate record of the force. In doing this, we omit most of the details and vary some of the parts, in order to make the explanation simpler and clearer, the object being merely to explain the principle.

The band-wheel a, fig. 114, (shown also in fig. 113,) revolves once for every yard of onward movement, as already stated. In doing so, it causes the arm $d\ c$, to vibrate backwards and forwards, on a pin at d; the connecting rod $b\ c$ being set near the circumference

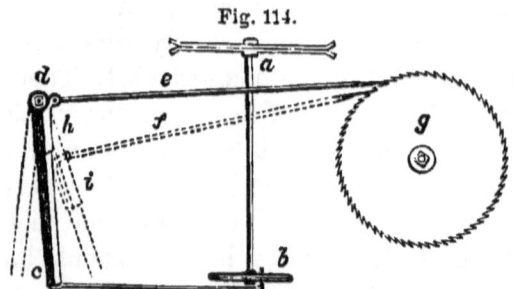

Fig. 114.

of the wheel b, this vibrating movement is shown by the dotted lines at f and i. The slide h moves on this vibrating rod, by being connected with the spiral spring already described, which indicates the force of the draught; the stronger the draught, the further this slide is moved toward c. When there is no draught at all, the rod e remains at the pivot d, and has no motion; but as the slide h is moved successively along the arm, this rod e is thrust backwards and forwards, more or less, according to the force of the draught. This thrusting movement turns the ratchet wheel g faster or slower as this force varies. A self-recording index is connected with this wheel by an arrangement similar to that already described for registering the distance.

This explanation shows the principle of the self-registering attachment, but in one respect it must be varied in order to be entirely accurate. The ratchet wheel must necessarily permit some play of the click or pawl, which would soon lead to serious error. This is wholly prevented by facing the wheel with India rubber, and causing the pawls to press this India rubber surface.

It will be observed that a movement of this wheel is made at every revolution of the band-wheel, or once in every yard; and in traveling a hundred yards, a hundred such movements are made. Every one of these may be different in amount from the others, yet the whole sum will be accurately measured.

It is absolutely essential that every part be finished with perfect workmanship, so that there may be no play or rattling of the teeth, producing loss of motion. Its measurements have been entirely satisfactory, although its records must necessarily vary with the condition of the cutting edge of plows, with the running order of mowing machines, the temper or sharpness of the knives, and the skill of the manager or driver.

A more general use of the dynamometer would doubtless result in important advantage to farmers as well as plow-makers. The trials which have been made, both in this country and in Europe, have proved that a great difference exists in plows, as to ease of draught,—some plows requiring a force more than fifty per cent greater than others, to turn a furrow of equal width and depth. Hence the farmer who employs the plow which runs most freely may accomplish as much by the use of two horses, as another can do by using one of hard draught by employing three horses.

DYNAMOMETER FOR ROTARY MOTION.

All these dynamometers apply only to simple, onward draught, as in plowing, drawing wagons, harrowing, etc. There is another, represented in fig. 115, of very ingenious but complex construction, which shows the force required in working any rotary machine, such as thrashers, straw-cutters, and mills, and showing, at the same time, the velocity, and recording the number of revolutions made.

The whole machine is supported by a cast-iron frame-

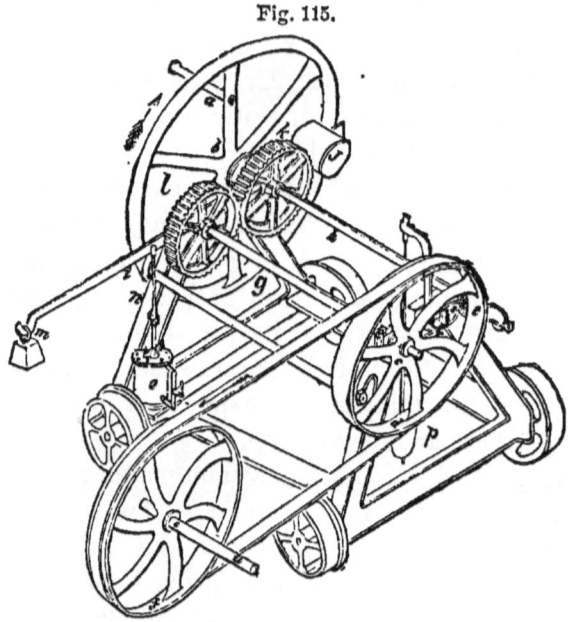

Fig. 115.

Dynamometer for measuring the force and velocity of thrashing-machines.

work, on four small wheels with flanges, like the wheels of rail-cars, that it may be conveniently run up on a temporary rail-way to the thrashing or other machine to be tried.

The band-wheel f, on the shaft e, is connected with the machine under trial, and the force is supposed, in this instance, to be applied by hand to the handle a, on the flywheel.

When the fly-wheel is turned in the direction shown by the arrow, it causes the two cog-wheels to revolve, and moves the band in the direction shown by the other arrow. Now, whatever force is required to turn the wheel *f*, connected with the machine under trial, must be overcome by a corresponding force applied to the handle *a*, because the wheel-work is so adjusted that this handle moves with the same velocity as the band on the band-wheels.

The wheel *f*, being connected by the band to the wheel *d*, which is on the same axis or shaft as the cog-wheel *l*, the resistance of the machine under trial tends to keep the cog-wheel *l* from turning, until enough force is applied to the handle *a*, to set the cog-wheel *k* in motion. Now the greater the resistance, the greater will be the power needed at the handle. This power, therefore, is measured accurately in the following manner:

The axle *g*, of the cog-wheel *l*, rests at its further end in an oblong hole or mortise, which allows it liberty to play, or rattle up and down within narrow limits. This same axle, *g*, passes through a hole in the lever *i* so that when it rattles up and down, it carries this lever up and down with it. The other part of the lever turns on the shaft *h* of the other cog-wheel.

Now when the man at the fly-wheel applies his force to the handle *a*, the resistance of the machine under trial causes the cog-wheel *l* to refuse to turn; consequently, his force, instead of turning it, lifts it up in the mortise, and raises the lever with it. As he increases his force against the handle, let weights be hung on the lever, until, at the very moment that the wheel begins to *revolve*, the weights shall be just heavy enough to keep the lever down in the mortise. *This weight, therefore, will measure the exact force needed to turn the machine:* the greater the resistance of the machine, the greater must be the weight.

There is another weight, *J*, used to balance the lever and cog-wheel *l*, while the machine is at rest, **or before** the force is applied to it, so that the weight at *m* shall represent the force truly. The weight *m* is, of course, to be multiplied by the power it exerts on the lever *i*, which should be graduated like the bar of a steelyard.

There are a few other parts of this dynamometer not yet described. One is the cylinder *o*, filled with oil, in which a perforated piston works, preventing the rapid vibration of the lever *i*, as the force varies, precisely similar to the cylinder of oil described in fig. 110, p. 100. Another part is the pendulum *p*, with the wheel *r*, which measures the time.

The use of this instrument has been already attended with some important results in detecting the great amount of friction existing in some thrashing-machines of high reputation, which has been found to amount, in certain cases, to more than one-half of the whole power applied. It is only by detecting so great a waste that we are enabled to take measures for its prevention.

CHAPTER VIII.

APPLICATION OF LABOR.

Most of the moving powers applied by the farmer to accomplish labor are the exertions of animal strength. A principal object of the preceding pages is to point out how this strength can be applied in the most economical manner, and to aid in the substitution of cheap horse-power for more costly human labor. It will doubtless contribute to the end to exhibit the relative efficiency of each, as well as the results of strength differently applied.

The amount of work which any machine is capable of performing is denoted by comparing this amount with the power of a single horse; hence the common expressions of twenty, or fifty, or a hundred horse-power engines. The strength of different horses varies greatly, but the expression, as commonly understood, indicates a force equivalent to raising or pressing with a force equal to 150 pounds 20 miles a day, at the rate of two and a half miles an hour. This is the same as 33,000 pounds raised one foot in one minute. The results of numerous experiments in different places give the actual power of the average of horses at somewhat less than this; and there is no doubt that, for most of the farm-horses of this country, the result would be considerably less. The power of a strong English draught-horse has been ascertained to be about 143 pounds for 22 miles a day, at $2\frac{3}{4}$ miles an hour. Many American horses are scarcely more than half as strong. The strength of a man, working at the best advantage, is estimated at one-fifth that of a horse.

As the speed of a horse increases, his strength of draught diminishes very rapidly, till at last he can move only his own weight. This is owing to three reasons: first, the load moves over a greater space in a given time, and if, for instance, the speed be doubled, half the load only can be carried with the same quantity of power, according to the law of virtual velocities; secondly, the horse has to carry the full weight of his body, whatever his speed may be, and the force expended for this purpose alone must, therefore, be doubled as the speed is doubled; thirdly, a very quick and unaccustomed motion of the muscles is in itself more fatiguing than the ordinary or natural velocity.

The following table shows the amount of labor a horse of average strength is capable of performing in a day at different degrees of speed, on canals, rail-roads, and on turnpikes. The force of draught is estimated at about 83 pounds. This is considerably less than the horse-power used in estimating the force of machinery, but it is as much as an ordinary horse can exert without being improperly fatigued with continued service:

Velocity per hour.	Duration of the day's work.	Work accomplished for one day, in tons, drawn one mile.		
Miles.	Hours.	On a canal.	On a rail-road.	On a turnpike.
$2\tfrac{1}{2}$	$11\tfrac{1}{2}$	520	115	14
3	8	243	97	12
$3\tfrac{1}{2}$	$5\tfrac{9}{10}$	153	82	10
4	$4\tfrac{1}{2}$	102	72	9
5	$2\tfrac{9}{10}$	52	57	7.2
6	2	30	48	6
7	$1\tfrac{1}{2}$	19	41	5.1
8	$1\tfrac{1}{8}$	12.8	36	4.5
9	$\tfrac{9}{10}$	9	32	4
10	$\tfrac{3}{4}$	6.6	28.8	3.6

From the preceding table it will be seen that a horse, at a moderate walk, will do more than four times as much work on a canal as on a rail-road; but the resistance of the water increases as the square of the velocity, and therefore when the speed reaches five miles an hour, the rail-road has the advantage of the canal. On the rail-road and turnpike the resistance is about the same, whether the speed be great or little, the chief loss with fast driving resulting from the increased difficulty with which the horse carries forward his own body, which weighs from 800 to 1200 pounds. The table also shows that when it becomes necessary to drive rapidly with a load, it should be continued but for a very short space of time; for a horse becomes as much fatigued in an hour, when drawing hard at ten miles an hour, as in twelve hours at two and a half miles an hour; because when a boat is driven through the water, to double its velocity not only requires that twice the amount of water should be moved or displaced in a given time, but it must be moved with twice the velocity, thus requiring a four-fold force.

The muscular formation of a horse is such that he will exert a considerably greater force when working horizontally than up a steep, inclined plane. On a level, a horse is as strong as five men, but up a steep hill he is less strong than three; for three men, carrying each 100 pounds, will ascend faster than a horse with 300 pounds. Hence the obvious waste of power in placing horses on steeply inclined tread-wheels or aprons. The better mode is to allow them to exert their force more nearly horizontally, by being attached to a fixed portion of the machine. For the same reason, the common opinion is erroneous that a horse can draw with less fatigue on an undulating than on a level road, by the alternations of ascent and descent calling different muscles into play, and relieving each in turn; for the same muscles are alike exerted on a level and on an ascent, only in the latter case the fatigue is much greater than the counterbalancing relief. Any person may convince himself of the truth on this subject by first using a loaded wheel-barrow or hand-cart for one day on a level, and for the next up and down a hill; bearing in mind, at the same time, that the human body is better fitted for climbing and descending than that of a horse.

A *draught-horse* can draw 1600 pounds 23 miles in a day, on a good common road, the weight of the carriage included. On the best plank-road he will draw more than twice as much.

A *man* of ordinary strength exerts a force of 30 pounds for 10 hours a day, with a velocity of $2\frac{1}{2}$ feet per second. He travels, without a load, on level ground, during $8\frac{1}{2}$ hours a day, at the rate of 3.7 miles an hour, $31\frac{1}{4}$ miles a day. He can carry 111 pounds 11 miles a day. He can carry in a wheel-barrow 150 pounds 10 miles a day.

Well-constructed machines for saving human labor by means of horse-labor, when encumbered with little friction, will be found to do about *five times* as much work for each horse as where the same work is performed by an equal number of men. For example: an active man will saw twice each stick of a cord of wood in a day.

Six horses, with a circular saw, driven by means of a good horse-power, will saw five times six, or thirty cords, working the same length of time. In this case the loss by friction is about equal to the additional force required for attendance on the machine.

Again: a man will cut with a cradle two acres of wheat in a day. A two-horse reaper should therefore cut, at the same rate, ten times two, or twenty acres. This has not yet been accomplished. We may hence infer that the machinery for reaping has been less perfected than for sawing wood. It should, however, be remembered, that great force is exerted, and for many hours in a day, in cutting wheat with a cradle, and therefore less than twenty acres a day may be regarded as the medium attainment of good reaping-machines when they shall become perfected.

Applying the same mode of estimate, a horse-cultivator will do the work of five men with hoes, and a two-horse plow the work of ten men with spades. A horse-rake accomplishes more than five men, because human force is not strongly exerted with the hand-rake.

In using different tools, the degree of force or pressure applied to them varies greatly with the mode in which the muscles are exerted. The following table gives the results of experiments with human strength, variously applied, for a short period:

	Force of the hands on the tool.	Force of the tool on the object.
With a drawing-knife	100 lbs.	100 lbs.
" a large auger, both hands	100 "	about 800 "
" a screw-driver, one hand	84 "	250 "
" a bench-vice handle	72 "	about 1000 "
" a windlass, with one hand	60 "	180 to 700 "
" a hand-saw	36 "	36 "
" a brace-bit, revolving	16 "	150 to 700 "
Twisting with thumb and fingers, button-screw, or small screw-driver	14 "	14 to 70 "

The force given in the last column will, of course, vary

with the degree of leverage applied; for example, the arms of an auger, when of a given length, act with a greater increase of power with a small size than with a large one. This degree of power may be calculated for an auger of any size, by considering the arms as a lever, the centre screw the fulcrum, and the cutting-blade as the weight to be moved. The same mode of estimate will apply to the vice-handle, the windlass, and the brace-bit.

Every one is aware that a heavy weight, as a pail of water, is easily lifted when the arm is extended downward, but with extreme difficulty when thrown out horizontally. In the latter case, the pail acts with a powerful leverage on the elbow and shoulder-joint. For this reason, all kinds of hand labor, with the arms pulling toward or pushing directly from the shoulders, are most easily performed, while a motion sidewise or at right angles to the arm is far less effective. Hence great strength is applied in rowing a boat or in using a drawing-knife, and but little strength in turning a brace-bit or working a dasher-churn. Hence, too, the reason that, in turning a grindstone, the pulling and thrusting part of the motion is more powerful than that through the other parts of the revolution. This also explains why two men, working at right angles to each other on a windlass, can raise seventy pounds more easily than one man can raise thirty pounds alone. This principle should be well understood in the construction or selection of all kinds of machines for hand labor.

CHAPTER IX.

MODELS OF MACHINES.

Serious errors might often be avoided, and sometimes gross impositions prevented, by understanding the difference between the working of a mere model, on a miniature scale, and the working of the full-sized machine. It is a common and mistaken opinion that a well-constructed model presents a perfect representation of the strength and mode of operation of the machine itself.

When we enlarge the size of any thing, the *strength* of each part is increased according to the square of the diameter of that part; that is, if the diameter is twice as great, then the strength will be four times as great; if the diameter is increased three times, then the strength will be nine times, and so on. But the *weight* increases at a still greater rate than the strength, or according to the *cube* of the diameter. Thus, if the diameter be doubled (the *shape* being similar), the weight will be eight times greater; if it be tripled, the weight will be twenty-seven times greater. Hence, the larger any part or machine is made, the less able it becomes to support the still greater increasing weight. If a model is made one-tenth the real size intended, then its different parts, when enlarged to full size, become one hundred times stronger, but they are a thousand times heavier, and so are all the weights or parts it has to sustain. All its parts would move ten times faster, which, added to their thousand-fold weight, would increase their inertia and momentum ten thousand times. For this reason, a model will often work perfectly when made on a small scale; but when enlarged, the parts become so much heavier, and their momentum so vastly greater, from the longer sweep of motion, as to fail entirely of success, or to become soon racked to pieces.

This same principle is illustrated in every part of the works of creation. The large species of spiders spin thicker webs, in comparison with their own diameter, than those spun by the smaller ones. Enlarge a gnat until its whole weight be equal to that of the eagle, and, great as that enlargement would be, its wing will scarcely have attained the thickness of writing-paper, and, instead of supporting the weight of the animal, would bend down from its own weight. The larger spiders rarely have legs so slender in form as the smaller ones; the form of the Shetland pony is quite different from that of the large cart-horse; and the cart-horse has a slenderer form than the elephant.

The common flea will leap two hundred times the length of its own body, and the remark has been sometimes made that a man equally agile, with his present size, would vault over the highest city-steeple, or across a river as wide as the Hudson at Albany. Now, if the flea were increased in size to that of a man, it would become a hundred thousand times stronger, but thirty million times heavier; that is, its weight would become three hundred times greater than its corresponding strength. Hence we may infer that the enlarged flea would be no more agile than a man; or that, if a man were proportionately reduced to the size of a flea, he could leap to as great a distance.

All this serves to illustrate in a striking manner the great difference in the working of models and of machines.

CHAPTER X.

CONSTRUCTION AND USE OF FARM IMPLEMENTS AND MACHINES—IMPLEMENTS FOR TILLAGE.

The application of mechanical principles in the structure of the simpler parts of implements and machines has been already treated of. It remains to examine more particularly those machines chiefly important to the farmer, and to show the application of these principles in their use and operation.

Farm implements and machines for working the soil should be, as far as possible, simple and not complex, because they mostly meet with an irregular resistance, consisting of hard and soft soil and stones variously mixed together. A locomotive is made up of many parts; but having a smooth surface to traverse, the machinery works uniformly and uninjured; but if in its progress it met with formidable obstructions and uneven resistance, it would be soon racked and beaten to pieces. Hence the long-continued and uniform success of the simple plow; as well as the failure of complex digging machines, unless worked exclusively in soils free from stone. A complex machine, that meets with an occasional severe obstruction, receives a blow like that of a sledge; and when this is repeated frequently, the probability is that some part will be bent, twisted, knocked out of place, or broken. If the machine be light, the chances are in its favor; but if heavy, its momentum is such that it can scarcely escape severe injury. If composed of many distinct parts, the derangement or breakage of one of these is sufficient to retard or put a stop to its working, and men and teams must stand idle till the mischief is repaired.

Hence, after the trial of the multitude of implements and machines, we fall back on those of the most simple form, other things being equal. The crow-bar has been employed from time immemorial, and it will not be likely to go out of use in our day. For simplicity nothing exceeds it. Spades, hoes, forks, etc., are of a similar character. The plow, although made up of parts, becomes a single thing when all are bolted and screwed together. For this reason, with its moderate weight, it moves through the soil with little difficulty—turning aside from obstructions, on account of its wedge form, when it cannot remove them. The harrow, although composed of many pieces, becomes a fixed solid frame, moving on through the soil as a single piece. So with the simpler cultivators. Contrast these with the ditching machine (Pratt's) considerably used some years ago, but ending in entire failure. It was ingeniously constructed and well-made, and when new and every part uninjured, worked admirably in some soils. But it was made up of many parts, and weighed nearly half a ton. These two facts fixed its doom. A complex machine, weighing half a ton, moving three to five feet per second, could not strike a large stone without a formidable jar; and continued repetitions of such blows bent and deranged the working parts. After using a while, these bent portions retarded its working; it must be frequently stopped, the horses become badly fatigued, and all the machines were finally thrown aside. This is a single example of what must always occur with the use of heavy complex machinery working in the soil. Mowing and reaping machines may seem to be exceptions. But mowers and reapers do not work in the soil or among stones; but operate on a soft, uniform, slightly resisting substance, made of the small stems of plants. Every farmer knows what becomes of them when they are repeatedly driven against obstructions by careless teamsters.

There is another formidable objection to complex machines—this is, their *cost*. Even with some of proved value, the expense is a serious item with moderate farmers. Mowers and reapers, $130; grain drills, $80 or $90; thrashing machines, $100 to $400; horse rakes, $45; hay tedders, $80 to $100; iron rollers, $50 to $100; and even some of the efficient new potato diggers are offered for not less than $100. Placing all these sums, and many others for necessary tools together, the whole will be found a large outlay — more economical by far, it is true, than doing without them; but greater simplicity and consequent cheapness, as well as durability, would facilitate progress in agricultural improvement. A single machine, Comstock's spader, is offered at $250—twenty times the price of the best cast-iron plow, and ten times that of the most finished steel plow. And yet it is applicable only to land free from stone.

The object of these remarks is to caution farmers against investing money in newly invented contrivances of high promise at first, which are liable to the objection pointed out; and also inventors and manufacturers themselves against engaging in enterprises having at hand golden promises, but with failure in the distance.

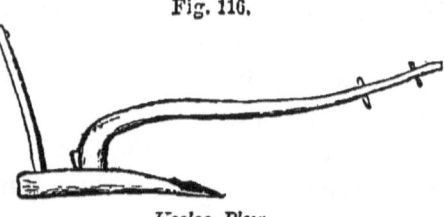

Fig. 116.

Kooloo Plow.

PLOWS.

The simplest plow, used probably in the earlier ages of the world, and found at the present day only among degraded nations, is the crooked limb of a tree, with a projecting point for tearing the surface of the earth. The above figure represents an improvement on the first rude implement, and is found at the present day in Northern

Fig. 117.

Moorish Plow.

India. Fig. 116 shows the Kooloo plow, consisting wholly of wood, except the iron point. Fig. 117 exhibits the implement now used in Morocco, which resembles the India plows, with the addition of a rude piece of timber as a mould-board. Both these perform very imperfect work, and have remained with little change for centuries, the owners not enjoying the benefit of agricultural read-

Fig. 118.

ing and intelligence. Fig. 118 is a step in advance, and represents a plow still used in some parts of Europe. In the less improved portions of Germany, the Baden plow,

Fig. 119.

Baden Plow.

represented by Fig. 119, is employed, and does not differ greatly from the "bull plow" commonly used in this country at the beginning of the present century. Great improvement has been made within the past fifty years, among others by the ingenuity and labors of Jethro Wood, and more recently by a great number of inventors and manufacturers in different parts of the country. Wood introduced the cast-iron plow into general and successful use, by cheapening its construction and perfecting its form,

Fig. 120.

Modern Improved Plow.

and others have made important improvements, including the steel mould-board now largely employed at the West.

Cast-iron plows have been generally used throughout the Eastern States; but for the peculiar soil of the West, it has been found absolutely necessary to use steel plows exclusively; and for the purpose of keeping them at all

Fig. 121.

Moline Plow.

times sharp for cutting the vegetable fibre and separating the parts of the soil readily, the practice is common to carry a large file or rasp for this purpose. These steel plows are made of plate previously rolled. They are becoming partially introduced also at the East, although in hard and gravelly soils the cast-iron mould-board is preferred by many, and regarded as even more durable. The steel plate plow is lighter than the cast-iron, but is more expensive. The accompanying figure

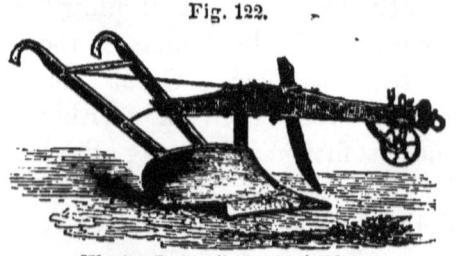

Fig. 122.

Woodruff & Allen's Steel Plow.

(Fig. 121,) represents the celebrated "Moline plow," made by Deere & Co., of Moline, Ill., one of the best and most extensively introduced among the Western steel implements; and Fig. 122 shows an excellent one of Eastern manufacture, made by Woodruff, Allen & Co., of Auburn,

N. Y. Good steel plows cost about double those made of cast-iron.

CHARACTER OF A GOOD PLOW.

Every good plow should possess two important qualities. The first relates to its working. It should be easily drawn through the soil, and run with uniform depth and steadiness. The second refers to the character of the work when completed. The inversion of the sod, especially if encumbered with vegetable growth, should be complete and perfect; and the mass of earth thus inverted should be left as thoroughly pulverized as practicable, instead of being laid over in a solid, unmoved mass. This is of the greatest importance on heavy soils, and is highly useful on those of a lighter character, except, it may be, clear sand or the lightest gravels. The harrow, at best, is an imperfect loosener; it pulverizes the surface, but its weight, and that of the team, press down the mass below. Whatever loosening, therefore, can be accomplished in plowing is a gain of vital importance.

THE CUTTING EDGE.

The point and cutting edge of the plow perform the first work in separating the furrow-slice from the land. It is important that this edge should not only do the work well, but with the greatest possible ease to the team. The force required to perform this cutting is greater than many suppose. The gardener who thrusts his sharp spade into the hard earth uses more force than afterwards in lifting and inverting the spit. We may hence infer that a large part of the power of the team is expended in severing the furrow-slice. This inference has been proved correct by the use of the dynamometer, in connection with carefully conducted experiments, which have shown the force usually expended for cutting off the side and bottom of the furrow-slice, in firm soils, to exceed all the rest of the force required to draw the plow. The point or share should therefore be kept sharp, and form as acute an angle as practicable, as shown in Fig. 123. Some plows which other-

Fig. 123.   Fig. 124.   Fig. 125.

wise work well are hard to draw because the edge, being made too thick or obtuse, raises the earth abruptly. Fig. 124.

Where stones or other obstructions exist in the soil, it is important that the line of the cutting edge form an acute angle with the land-side, or, in other words, that it form a sharp wedge, (Fig. 125.) It will then crowd these obstructions aside, and pass them with greater ease than when formed more obtuse, as shown in Fig. 126, for the same reason that a sharp boat moves more freely through the water than one which is blunt or obtuse. The gardener or ditcher proves this advantage when he thrusts a sharp-pointed shovel, Fig. 127, more easily through stony or gravelly soil, than one with a square edge. (Fig. 128.)

Fig. 126.   Fig. 127.

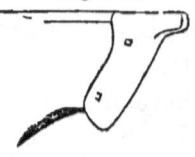

Fig. 128.

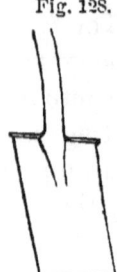

But when the soil is free from stones, or obstructions, or is filled with small roots which the plow should cut off, as in the Western prairies, the sharpness of the edge is more important than its form; and hence the reason that the use of the rasp or file becomes necessary in the field, to keep a sharp cutting edge at all times on the share.

NOTE.—It has been shown in the Report of the Trial of Plows at Utica, that so far as yet determined by experiment in England, about thirty-five per cent of the whole required draught is expended in overcoming the friction of the implement on its bottom and sides, about fifty-five for cutting the furrow-slice, and only about ten per cent for turning the sod. Hence the exclusive attention formerly given to forming the mould-board, as a means of reducing the draught, should have been directed more to lessening the force required for cutting the hard soil.

These experiments, however, do not appear to have been entirely satisfactory, especially for the light plows of this country; and it may be interesting to test their accuracy by calculation. The average weight of hard earth is about 125 lbs. per cubic foot; and the average draught of plows at the trial near Albany in 1850 was about 400 lbs. for a furrow-slice a foot wide and six inches deep. If a team in turning such sod moves two miles an hour, it raises a slice three feet long, equal to a cubic foot and a half (weighing 187 lbs.,) six inches each second—which would be the same as raising 31 lbs. three feet per second, which is the velocity of the plow. The mere force required to turn the sod, not estimating friction, would therefore be only one-thirteenth of the 400 lbs. of draught force. But the friction of dry earth on smooth iron is never less than one-half its weight; and if the earth is slightly plastic, its friction often is equal to, and sometimes exceeds, its weight. Taking the smallest amount, the friction on the mould-board would be equal to half the weight of the portion of sod resting on the mould-board, or about 31 lbs. This increased weight would also add equally to the friction of the sole of the plow, or 31 lbs. more—making the whole friction 62 lbs.; which added to the weight of the sod would amount to 93 lbs. —or more than one-fifth of the whole draught.

To ascertain the amount of friction, suppose the plow weighs 100 lbs. Half its weight would be 50 lbs., the friction on the sole of the plow. The friction of the sides would vary greatly with plows, being very small with those having a perfect centre-draught, or with no tendency to press against the land on the left. The whole friction and force for lifting the sod would therefore be about 150 lbs.; leaving 250 lbs. as the force for cutting the slice. A very easy running plow would leave a much smaller force—some as low as 200 lbs.

This estimate is liable to great variation. A wet and clayey soil would double the friction; a very hard piece of ground would add much to the force required for cutting the slice; if loose, the force would be comparatively small; or if quite moist, this force would be also much diminished; while the great difference in the draught of plows would vary the results still farther. The estimate, however, for soil dry enough to be friable, and of medium tenacity, is probably not far from correct, for plowing in this country—showing that most of the force required is for the act of cutting, and indicating the importance of giving special attention to the cutting edge.

THE MOULD-BOARD.

A prominent difference between good and bad plows results from the form of the mould-board. To understand the best form, it must be observed that the slice is first cut by the forward edge of the plow, and then one side is gradually raised until it is turned completely over, or bottom side up. To do this, the mould-board must combine the two properties of the wedge and the screw.

The position of the furrow-slice, from the time it is first cut until completely inverted, may be represented by placing a leather strap flat upon a table, and then, while

Fig. 129.

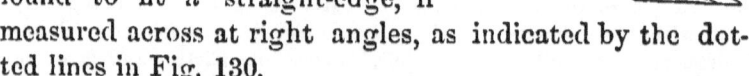

holding one end, turning over the other, so as to bring that also flat upon the table, as in Fig. 129.

Now, if the sole object were merely to invert the sod, the mould-board might have just such a shape as to fit the furrow-slice while in the act of turning over, or resemble precisely the twist of this leather strap. All the parts of this screw will be found to fit a straight-edge, if

Fig. 130.

measured across at right angles, as indicated by the dotted lines in Fig. 130.

But there are two objections to this form in practice. The first is that the sod is laid over smoothly and unbroken, and without being at all pulverized. On heavy and hard soils this is a serious fault. The other objection is that the sod is elevated as rapidly at the first movement, when its weight is considerable, as just before falling, when its pressure on the mould-board is slight. These difficulties are in part removed by giving the mould-board a shorter twist towards its rear. This form is distinctly shown in the figure of Holbrook's Stubble Plow, on a future page; and it contributes largely to that crumbling movement of the sod, so important for effecting pulverization.

The mould-board of a plow is capable of an almost infinite variety of forms, and the multitude of inventors have each adopted a different one. Some have made their selections by repeated random trials; while others, among whom Thomas Jefferson was the first, devised a series of straight lines, mathematically arranged, by which uniformity was given to the shape. The limits of this work preclude a full explanation. Many modifications in combining lines have been adopted, the most successful of which is that of Ex-Governor Holbrook, of Vermont, whose plows made according to these rules have performed admirably. It is less essential that farmers generally should understand these mathematical principles, provided they find a plow that will do good work; because, as already shown, the form of the mould-board has comparatively little to do with the required draught of the team.

Fig. 131.

It will be readily understood, however, that more force will be needed for drawing a short or blunt plow, like Fig. 131, than one in the form of a longer wedge, as in fig. 132, the latter, like a sharp boat in water, moving more easily. Care must be taken, however, that this slender wedge be not too long, else the friction of the sod on the extended surface may overbalance the advantage. The cutting part of the plow may be improperly formed like the square end of a chisel, and the sod may slide backward on a rise, with a very slight turn,

Fig. 132.

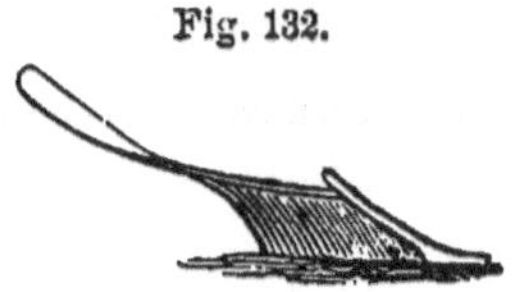

until elevated to a considerable height before inversion; this must require more force of the team, and make the plow hard to hold, on account of the side pressure. The character of this kind of plow may be quickly perceived by simply examining the mould-board after use; the scratches, instead of passing around horizontally, as they should do, are seen to shoot upward across the face and disappear at the top.

Instead of this form, the point should be long and acute, and the mould-board so shaped as to begin to raise the left side of the sod the moment it is cut, and before the right side is yet reached by the cutting edge. This turning motion being continued, the sod is inverted by being scarcely lifted from its bed; and the pressure which turns it being opposite to the pressure of the land-side, an equilibrium of these two pressures is maintained, and the plowman is not compelled to bear constantly to the right to keep the plow in its place.

Fig. 133.

Holbrook's Stubble Plow, or Deep Tiller.

There is, however, an exception, in deep or trench plowing, where it becomes necessary to throw the earth from the bottom of a furrow to the top of the inverted sod. A plow of this kind is represented in Fig. 133, which shows Holbrook's deep tiller for stubble land, capable of plowing a furrow a foot deep, and elevating the earth, which passes lengthwise over the mould-board. A similar form must be adopted for the rear mould-board of the Double Michigan Plow, so that the lower earth of the furrow may be thrown on the sod inverted by the first or skim-plow.

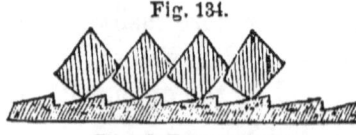

Fig. 134.

Crested Furrow-slicer.

The share should also be so placed as to cut the slice at equal thicknesses on both sides. Some plows are made so as to cut deepest on the land-side, forming a sort of saw-teeth section to the unmoved earth below, and leaving what is termed crested or acute ridges at the top. (Fig. 134.) Such plowing requires as much force in cutting the slice, and nearly as much in turning it over, as when level furrows are made, and should therefore be avoided. The same result is produced when

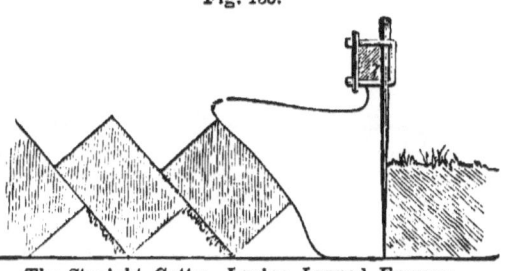

Fig. 135.

The Straight Cutter, Laying Lapped Furrows.

the plow is improperly gauged, and the plowman is compelled to press the handles to the left, to keep it from running too much to land.

On heavy or clay soils, it is sometimes desirable to place inverted sod in an inclined or lapping position, in order to give more exposure to the crumbling action of the weather, and to effect better drainage beneath. Fig. 135 is a section of these lapped furrows. In order to be equally inclined on both sides, their thickness must be precisely two-thirds their breadth; that is, if the plow runs eight inches deep, the slices should be twelve inches wide. This mode of plowing is controlled by the position of the cutter, which should be very nearly upright, as shown in Fig. 135. It has been justly remarked that the cutter to a plow (Fig. 136,) is almost as important as the rudder to a ship, and if its position be altered, as shown in Fig. 137, so as to cut under the sod, the furrows will cease to be lapped and will lie flat. This position is desirable in light or loose soils where exposure to the action of the

Fig. 136.

The Cutter.

air is not desirable, and where it becomes more important to bury completely all vegetable growth on the surface. If furrows are cut wider in proportion to their depth, they will be more likely to be laid flat. For example, if the plowing is six inches deep, and the furrows are a foot wide, the sod will generally dispose itself in a horizontal or flat position, and this result will be the more certainly secured by giving the form to the cutter already described. Lapping the furrows is the common practice in England, but is less necessary for this country, where the moisture of rains dries more quickly, and the severer frosts effect a ready pulverization; and especially is the practice less needed in thoroughly drained land.

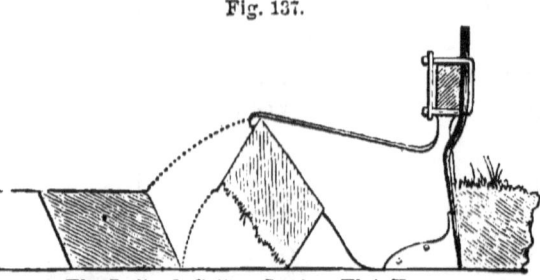

Fig. 137.

The Inclined Cutter, Laying Flat Furrows.

The Committee for the trial of implements, appointed by the New York State Agricultural Society, enumerated the following desirable qualities in plows, which every farmer may find useful to examine when he is about to purchase. 1. Pulverizing power. 2. Non-liability to choke in stubble. 3. Lightness of draught, considered in connection with pulverizing power. 4. Ease of holding. 5. Durability. 6. Cheapness. 7. Excellence of mechanical work. 8. Excellence of material. 9. Thorough inversion and burial of weeds. 10. Even distribution of wear. 11. Regularity or trueness of turning and carrying the furrow-slice in sod.

OPERATION OF PLOWING.

The expert plowman so adjusts his implement that it will cut a furrow of just such width and thickness as may be done with the least draught to the team, and the least exertion to himself. "To secure this end," says Todd, "the team is hitched as close to the plow as it can be and not have the whiffle-trees hit their heels in turning at the corners. As the length of the traces is increased, in plowing, the draught increases. Now put the connecting ring, or link, or dial clevis, at the end of the beam, in the lowest notch; and if it will not run deep enough, raise it another notch at a time until it will run just deep enough. Now alter the clevis from right to left, or from left to right, as may be necessary, until the plow will cut a furrow-slice just wide enough to turn it over well. If the plow crowds the furrow-slice without turning it over, it shows that the furrow-slice is too narrow for its depth; and the plow must be adjusted to cut a wider slice. On the contrary, if the plowman is obliged to constantly push the furrow-slice over with his foot, if the ground he is plowing be very smooth and even, it shows that there is an imperfection or fault somewhere. Sometimes by adjusting a plow to run an inch deeper, it will do very bad work. And sometimes it is necessary to adjust it to cut a little wider, or a little narrower, before it will cut the furrow-slice as well as it ought to be cut. When a good plow is correctly adjusted, it will glide along, where there are no obstructions, without being held, for many rods. When a plow is constantly inclined to fall over either way, and the plowman must hold it up all the while, to keep it erect, there is either an imperfection in the construction of the plow, or it is not adjusted correctly. When a plow "*tips up behind*," and does not keep down flat on its sole, or when it seems to run all on the point, either the point is too blunt, or is worn off too much on the *under* side, or there is not "*dip enough*"— pitching of the plow downwards—to the point. Sometimes I have found that a plow could not be adjusted by the clevis so correctly as all the parts were arranged; and

6*

that by shortening the traces or draught chain, or giving them a little more length, it would run like another plow. When a plow is adjusted to run just right, as the point wears off it is necessary many times to give a little more length to the draught chains, or to adjust it with the clevis to run a little deeper. It is sometimes impossible to adjust a plow to run just right with the style of clevis which is on the end of the beam. The arrangement ought always to be such that the draught can be adjusted half an inch at a time, either up or down, or to the right or left. Then if the beam of the plow stands as it should, so that the most correct line of draught *will cut the end of the beam*, it can be most correctly adjusted in a few seconds.

"To make a plow run deeper, raise the connecting point at the end of the beam one or more notches higher in the clevis; or lengthen the draught chains. To make it run more shallow, lower the draught a notch or more in the clevis; or shorten the draught chains; or, which should never be done, shorten the back-bands or hip-straps of the harness. To make a plow take a *wider* furrow-slice, carry the connecting point one or more notches in the clevis to the right hand. A notch or two to the left hand will make a plow cut a narrower furrow-slice. Or, which is seldom allowable, a plow may be made to run more *shallow* by putting the gauge-wheel lower, so as to raise the end of the beam. And a plow may be made to cut a narrower furrow-slice by carrying the handles to the left hand, or wider by carrying and holding them to the right, beyond an erect position; neither of which is allowable, except for a temporary purpose."

FAST AND SLOW PLOWING.

It has already been shown in the chapter on Friction, that the resistance is scarcely increased by velocity, when one body slides over another. The same rule, nearly, appears to apply to force required for cutting the earth, And as the friction of the plow and the force exerted in cutting the earth have been found to be the greater part of the whole draught, repeated experiments by the dynamometer have proved that but little increased resistance, as an average, occurs when a plow is drawn with increased velocity; the only additional power being that of doing more work in a given time. For example, if a force of 400 lbs. be required to draw a plow, whether at two or at four miles an hour, then twice as much power only is needed to plow an hour at four miles, as at two miles per hour. In other words, no more actual force in amount is necessary in most instances for a team to plow an acre in four hours at the faster speed than in eight hours at the slower. Hence the importance on the score of economy in time, of employing horses that have a naturally rapid gait, provided they possess full strength to overcome the required draught with ease. Fast plowing, however, is better adapted to stubble land than sod.

THE DOUBLE MICHIGAN PLOW.

The Double Michigan, called also the sod and subsoil plow, possesses some important advantages. The forward or skim plow pares off a sod a few inches in thickness, and inverts it into the bottom of the previous furrow. The second or main plow follows, and throws up the lower soil, completely burying the inverted sod and giving a loose, mellow surface to the field. This forms an excellent preparation for all crops, particularly carrots and other roots, which grow best in a deep, loose bed of earth; and where a portion of the subsoil improves the top-soil by being mixed with it, a permanent advantage results. A greater depth may be attained by the use of this double plow than with one having a single mould-board, in sod ground, because the inversion will be complete even if the width of the furrow is only one-half the depth. But with a single plow, the width must be considerably greater than the depth, or the sod will be thrown on its side or edge and cannot be inverted. There is one disadvantage, however, in the use of the double plow. A greater force is required to make two cuts in the soil, one above the other, than one cut with a single share.* For this reason more force must be used to plow a field to a given depth, say one foot, with the double than with the single plow. But the single plow, in order to reach this depth, would require to be so large and to turn so wide a furrow that no ordinary amount of team could be had to do the work. And in addition to this difficulty the inverted surface would not be so well pulverized as by the use of the double plow.

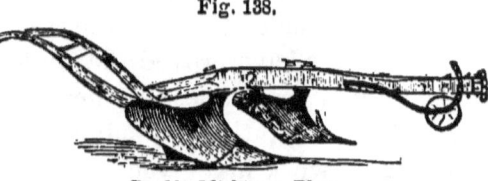

Fig. 138.

Double Michigan Plow.

THE SIDE-HILL PLOW.

Side-hill or Swivel plows are well known, and are so constructed as to throw the furrow-slice down hill, whichever way the team may be passing. The mould-board is turned to the right and left alternately for this purpose, the right-hand horse walking in the furrow in one direction, and the left-hand horse in the other. This plow is sometimes used for level land when it becomes desirable to avoid dead furrows and ridges, without plowing around the field. Fig. 139 represents the swivel plow manufac-

---

* This result has been proved by the use of the dynamometer; which has also shown that a greater amount of earth, in cubic feet, may be turned over with a deep-running plow than with a shallow one, as there is less force expended in cutting the slice when compared with the whole bulk—provided the soil is nearly uniform in hardness at different depths.

THE SIDE-HILL PLOW.

tured by F. F. Holbrook & Co., Boston, one of the best in use, and particularly valuable for its thorough pulverization of the soil. One-half of the double mould-board shown in the cut is used for throwing the furrow to the right, and the other half to the left—the change being effected by passing it under the plow with a single movement and hooking it in place.

Fig. 139.
Holbrook's Patent Swivel Plow.

Holbrook's Swivel or Side-hill Plow.

THE SUBSOIL PLOW.

When the common two-horse plow alone is used by farmers, it pulverizes the soil only a few inches in depth,

Fig. 140.

Subsoil plowing in the furrow of a common plow.

and its own weight and the tread of the horses on the bottom of the furrow gradually form a hard crust at that depth, through which the roots of plants and the moisture of rains do not easily penetrate. Hence the roots have only a few inches of good soil on the surface of the earth for their support and nourishment; and when heavy rains fall, the shallow bed of mellow earth is soaked and injured by surplus water. Again, in time of drought, this shallow bed of moisture is soon evaporated, and the plants suffer in consequence.

But, on the other hand, when the soil is made deep, it absorbs, like a sponge, all the rains that fall, and gradually gives off the moisture as it is wanted during hot and dry seasons. For this reason, deep soils are not so easily injured by excessive wetness, or by extreme drought, as shallow ones. In addition to this advantage, they allow a deeper range for the roots in search of nourishment.

Soils are deepened by *trench-plowing* and by *subsoiling*. In trench-plowing, the common plow with a mould-board is made to enter the earth to an unusual depth, and to throw up a portion of the subsoil, covering with it the top-soil which is thrown under. A subsoil plow, on the contrary, only *loosens* the subsoil, but does not lift it to the surface.

The Double Michigan Plow, just described, is strictly a trench-plow, and is one of the best implements for this purpose.

When the subsoil is of such a character that its mixture with the surface tends to render the whole richer, trench-plowing is best; but when of a more sterile character, it should be only loosened with the subsoil plow, and more cautiously intermixed with the richer portion above.

It often happens that the subsoil plow is very useful in loosening the soil for the purpose of allowing the trench-plow to run more freely through it.

The operation of the subsoil plow is shown in fig. 140.

In using the subsoil plow the less the earth is raised, provided it is well broken to pieces, the easier will be the draught. The part which moves under the soil and performs this loosening is of course in the form of a wedge. If the subsoil is dry, hard, and not adhesive, a long and acute wedge will run most easily; but if the subsoil is stony, a shorter wedge will succeed better. For general purposes it should therefore be of medium length.

Different modes of connecting this wedge to the beam above have been adopted, each possessing its peculiar advantages. Fig. 141 represents a subsoil plow with a single, broad, upright shank, cutting like a wedge, with double edges as well as double points, and capable of being reversed when it becomes worn. In light or gravelly soils this plow runs well; but where the earth is adhesive and rather moist, the friction of the two faces of this shank in pressing the

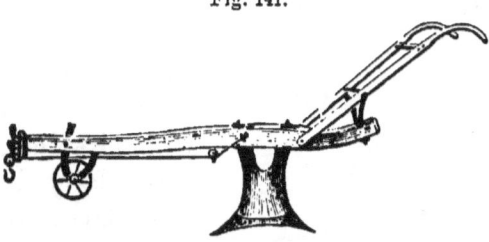
Broad-shank Subsoiler.

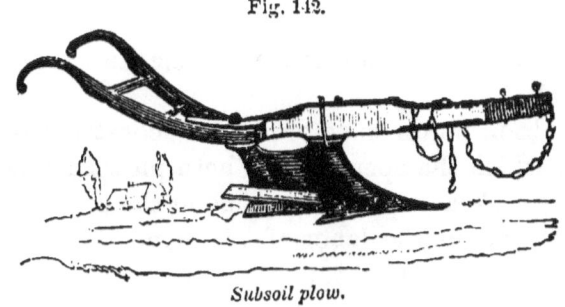
Subsoil plow.

compact soil apart becomes enormous, amounting in some cases to more than triple the force required to loosen the soil below. This plow is therefore not to be recommended for general use. The objection is in a great measure obviated in the plow shown in fig. 142,

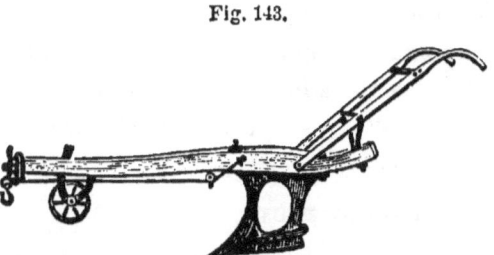
Two-shanked Subsoiler.

where the forward portion of the broad plate is made thicker than the rest. The friction is still further lessened by employing two narrow shanks, as in fig. 143.

Another improvement for lessening friction might be made by using narrow bars of iron or steel, braced and connected as shown in fig. 144. The ditching plow, exhibited in fig. 147, is similar in the construction of this part, and it has been found to work well for subsoiling, particularly in stony land. If the subsoil happens to be filled with roots, the interstices in these plows sometimes become choked—a difficulty, however, which rarely occurs. In such cases it may be better to employ the plow represented by fig. 141.

Fig. 144.

Brace-shank Subsoiler.

New subsoil plows have been lately constructed at the West, by which the operations of both plows are performed at once. A saving is thus made in the expense of the implement and in the labor of one man. In one, known as the Nichols' plow, a flat, triangular blade runs a few inches below the common plow; in Wheatley's, a narrow blade bent like the letter U beneath the plow performs the work.

The benefit of subsoiling will last three or four years; but it is of great importance that land be well underdrained, for if the earth becomes heavily soaked with water, it settles down into one compact mass, and the advantages of the operation are lost.

THE PARING PLOW

consists merely of a flat blade, which runs beneath the surface, shaving off the roots, but not moving the soil (fig. 145). A shield, shown in the cut, is placed beneath the beam, to regulate the depth of the cutting blade. It is used in cutting turf for burning, and for destroying thistles and other deep-rooted weeds. When made light for a single horse, it is sometimes used advantageously for cutting the grass and weeds between rows of badly tilled corn. A two-horse paring plow has been constructed, in which the depth of cutting is accurately regulated by wheels placed on an axle, like those of a cart. The cast-

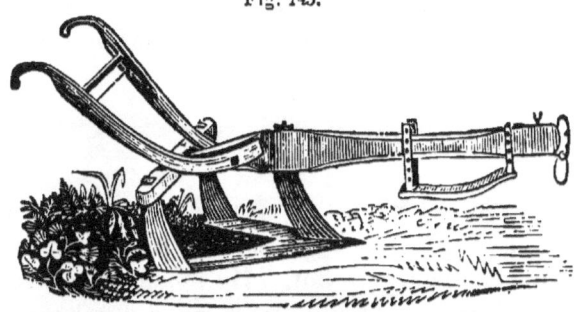

Paring plow.

iron blade, which cuts about three feet wide, is raised or depressed by means of screws passing through the axle. Its chief utility is in destroying grass and weeds before the sowing of broadcast crops.

THE GANG PLOW

consists of three or four small mould-boards placed side by side (fig. 146), and is used for shallow plowing, or for

Gang plow.

burying manure or seed on inverted sod, without disturbing the turf beneath. In those of the best construction, the depth is regulated by wheels, and the breadth of the furrows by turning the cross-beam more or less obliquely, by means of a fixed contrivance for this purpose. The gang plow is liable to become impeded or clogged by stubble, coarse manure, or weeds, and has not come into extensive use.

DITCHING PLOWS.

In most localities where tile drains are made, two-thirds of the labor of cutting is loosening the earth with the pick, before shoveling it out. By means of the ditching plow this laborious work is performed by horses. One span, with a good plow made for this purpose, will loosen the subsoil fast enough for eight or ten men shoveling, and cutting about 100 rods 3 ft. deep in a day; or an hour or two each day with the plow will keep two men at work. If the subsoil is very hard, this work should be done early in summer. The implement is drawn by two horses, attached to the ends of a main whiffle-tree about seven feet long, one walking on each side of the ditch. From one to three times passing will loosen the subsoil five to eight inches, which is then thrown out by narrow shovels, on both sides, so that it may be easily returned after the tile is laid, by means of a common plow drawn by the long whiffle-tree before mentioned.

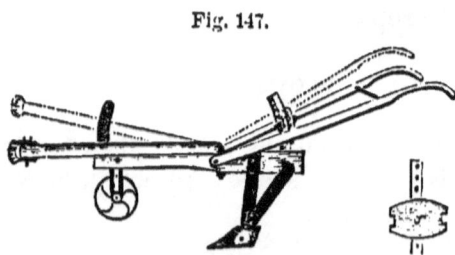

Fig. 147.

Adjustable Ditching Plow.

There are several modifications of the ditching plow, all accomplishing the same end. *The adjustable ditching plow,* (fig. 147,) admits of so great a change in the height of the beam and handles, that it may be run down in the bottom of a ditch to a depth of four feet. It is, perhaps, the best implement of the kind for all purposes and soils. The movable portion of the beam is attached to the fixed beam by a stout loop and staple, and rises on a cast-iron arc, which passes through it, as shown by the dotted lines. The handles rise on a stiff, *wooden* arc, (as the dotted lines exhibit,) a piece of thick plank, shown in the small figure on the right, being placed between the handles and fastened to them, to render them more firm and steady. The iron work, although light, is braced so as to impart great strength and security. The point is screwed on separately, and is nearly the only part that wears by use.

This ditching plow may be used for common subsoiling, the shortness of the share rendering it especially adapted to stony land.

Several ditching machines have been constructed for performing the entire operation of cutting the earth and throwing it out, but nearly all of them are too complex for common use. Except in land entirely free from stone, some of their many parts are liable to become bent or injured by use, and a very slight derangement of this kind renders them partly or entirely useless. Any ditching machine, therefore, to work well among stone, must be simple and strong, so as to withstand the frequent shocks met with in overcoming obstructions in the soil.

MOLE PLOW.

The Mole Plow has a wooden beam, sheathed with iron on the lower side, which moves close to the ground, below which a thin, broad coulter extends downward, and to the lower end of this coulter a sharp iron cylinder is attached. This moves horizontally, point foremost, through the soil, producing a hollow channel beneath the plow for the escape of the water, the only trace on the surface being a narrow slit left by the coulter. It is dragged forward by means of a chain and capstan worked by a horse, the machine itself being fixed with strong iron anchors. This mode of draining is only adapted to clay soil, free from stone, and although cheaply performed, has been little used since the introduction of tile-draining.

APPENDAGES TO THE PLOW.

WHEEL COULTERS.—In soils free from stones and coarse gravel, and especially on the Western prairies, wheel coulters are found to answer a good purpose, cutting through the turf and roots of grass with great ease, and making a smoother slice than the common cutter. But where stones and other obstructions exist, it is necessary to use the simpler, single blade coulter. A good representation of the wheel coulter is seen on the figure of the Moline Plow, on an early page of this chapter.

WEED-HOOK AND CHAIN.—In turning under large weeds, grass, or other tall vegetable growth, two modes are adopted. One is the use of the weed hook represented in the annexed cut; and the other is that of a chain. The weed-hook has been long known, and is made in various

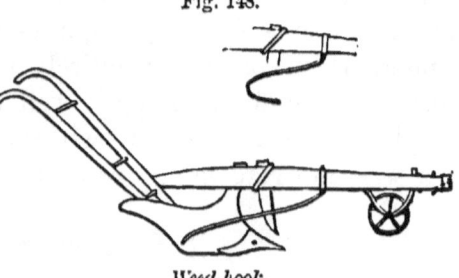

Fig. 148.

Weed-hook.

forms. Sometimes it is bent in the form of a bow with the lower point projecting forward, as in the upper figure; another form is like that shown in the lower cut, pointing backwards. This is less liable to be caught by obstructions. The weed-hook operates on the principle of bending the tall growth forward and prostrate, so that the turning sod completely buries it. The same object is attained by the use of a heavy chain; and different modes are used for attaching it to the plow. One of the simplest is to fasten one end to the right-hand portion of the main whiffle-tree, and the other to the right handle. In another mode, the chain forms a loop. All these modes of burying vegetable growth are important in turning under clover and other green crops.

The weed-hook is usually made of round rod-iron, stiff enough to perform its work, and to possess some spring when it meets with obstructions. Those not accustomed to its use may adjust its position by bending it, until it performs satisfactorily. It is secured to the plow-beam by placing the forward end in a small groove cut lengthwise in the under side of the beam, passing a band over it, and wedging until properly secured. Lighter and more perfect weed-hooks may be made of steel rod, similar to that used for rake teeth; they will bend back on meeting obstructions, and spring again into position. Such weed-hooks should be made and sold with the other appendages of plows, now that the inversion of green clover for manure has become an essential part of good farming. Sometimes the weed-hook is made to extend at right angles to the plow-beam, curving outwards and downwards. This form requires greater stiffness, and small bar-iron is used.

No plow will cover weeds or other growth two or three feet high; but by the use of this hook, the whole is laid completely under the surface.

REGULATING WHEEL.—It has long been a question with plow men whether the wheel under the beam for regulating depth is really a disadvantage or a benefit. It is fully shown in the able Report by J. Stanton Gould, of the Trial of Plows at Utica, drawn from accurate experiments, that the wheel not only gives better plowing with moderate skill, but that it slightly lessens the draught. Uniformity in the depth of the slice is preserved, without constant vigilance on the part of the attendant; and this uniformity, by preventing uneven running, lessens the aggregate amount of draught. It is, however, quite important that the wheel sustain little or no pressure; for as soon as the beam bears upon it, the line of draught becomes crooked at the expense of the team. These facts were established by careful experiments with the dynamometer.

PULVERIZERS.

The fine pulverization of the earth, for the ready extension of the roots of plants, for the action of air on the soil, for the retention of moisture, and for the thorough intermixture of manure, is of great importance to the farmer. It is but partially accomplished by the plow, which crumbles the soil only so far as may be done by the act of turning it over. Hence additional implements are needed for this purpose, among which are the *harrow*, the *cultivator*, and the *clod-crusher*.

HARROWS.

The *Brush-harrow*, the original and rudest form of the implement, and still used for covering grass seed, as often made, is a poor implement. The most projecting limbs are cut partly off, that all may lie flat, but it often happens that the projecting angles of the larger branches plow into the ground and make deep furrows. This may be prevented by a careful selection of the small tree which forms the brush, or still better by constructing a simple rough plank frame, so that any quantity of short brush may be placed between two pieces of plank, to admit the tops of the brush to incline downwards and backwards, being held in place by a few spikes or bolts. Fig. 149.

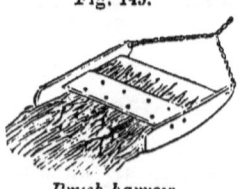
Fig. 149.
Brush-harrow.

The *Geddes Harrow* is one of the best in use for rough or uneven land. The teeth being situated considerably back of the point of draught, its motion is even and steady, and easy for the team. In consequence of its wedge-form, it passes obstructions more readily. The center or draught-rod forms a set of hinges, by which it becomes adapted to uneven ground, or by which it may be easily lifted to discharge weeds, roots, or other obstructions. Or it may be doubled back, and carried easily in a wagon. The accompanying figure (fig. 150) renders its construction intelligible, without further description. To prevent its rising in the middle, as it has been found to do when the draught traces are as short as easy draught requires, the chain is attached to the bar on each side, as shown in fig. 151.

Fig. 150.

Geddes Harrow.

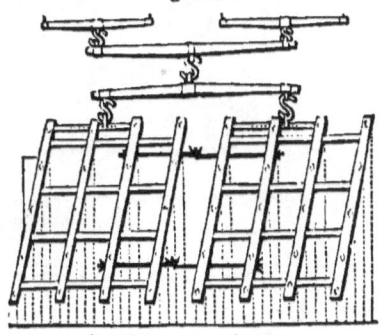

Fig. 151.

Fig. 152.

Scotch or square harrow.

The *Square Harrow* admits of a larger number of teeth, and when made in the best form, effects thorough pulverization on smooth land, free from obstructions. A modification known as the Scotch harrow, represented in fig. 152, has forty teeth, inserted in such a manner that each tooth forms a separate track, as shown by the dotted lines. The hinges, as in all square harrows, enable it to fit a rolling or uneven surface, and it may be folded for carrying in a cart or wagon. For the fine pulverization of a smooth surface, a still greater number of teeth has been found to answer an excellent purpose, leaving the soil almost as smooth as a garden bed. Tough and sound timber, only two inches square, is used for the frame, and the teeth are five-eighths of an inch square.

The *Morgan Harrow* is an improvement of the Scotch implement, slots being made in the hinges, so that each of the two portions is capable of playing freely up and down, as the surface varies, and rendering the rear teeth less liable to follow in the track of the preceding. The draught-iron is made to slide on an iron arc, so that the lines formed by the teeth are controlled at pleasure. It is converted into a broadcast cultivator by inserting flat teeth, the flat portion below being the same in width as above, and pointing slightly forwards. These teeth pulverize the soil deeply and thoroughly. They are successfully used for digging potatoes, operating like a large number of potato-hooks, drawn by horses.

The *Norwegian Harrow* (fig. 153) is a new machine for

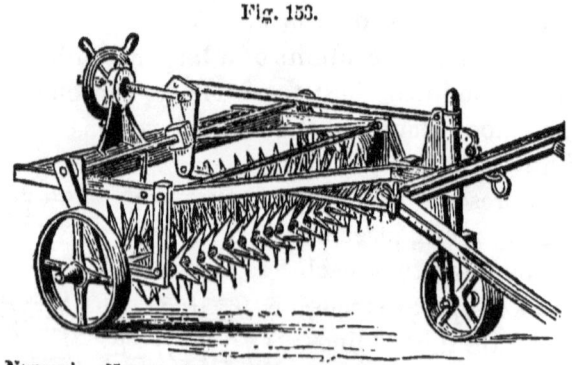

Fig. 153.

Norwegian Harrow, kept from clogging by two cylinders of teeth playing into each other.

pulverizing the soil, which performs the work in a very perfect manner, by turning up, instead of packing down, the earth. Two rows of star-shaped tines play into each other, and produce a complete self-cleaning action, preventing clogging even in quite adhesive soils. Its complex character and cost have prevented its coming into more general use.

Shares' Harrow (fig. 154) is the most perfect of all implements for pulverizing the freshly inverted surface of sward land, to a depth two or three times as great as the common harrow can effect. The teeth being sharp, flat blades, cut with great efficiency; and as they slope like a sled-runner, they pass over the sod, and instead of tearing it up like the common harrow or gang-plow, they tend to keep it down, and in its place, while the upper surface of the sod is sliced up and torn into a fine, mellow soil. The price of Shares' harrow is about $20, but if furnished with steel teeth, as it should be, it would cost more.

Fig. 154.
Shares' Harrow.

CULTIVATORS.

The *Cultivator* or *Horse-hoe* is used for loosening and pulverizing the soil among drilled crops, and for cutting and destroying weeds. A usual form is shown in fig. 155, which represents Holbrook's, one of the best of its kind. The wheel in front regulates the depth; the sides may be expanded or contracted sufficiently to vary the width from fifteen to thirty-six inches; they are reversible, so that the soil may be thrown from or towards the row; and the frame is high enough to prevent clogging with

Fig. 155.
Holbrook's Horse-hoe or Cultivator.

weeds, stubble, or manure. Various forms of teeth are used, according to the nature of the work, and they are made of steel or cast-iron. The steel teeth, represented in fig. 152, are well adapted for cultivating the rows of Indian corn and other hoed crops, where the soil is al-

Fig. 156.

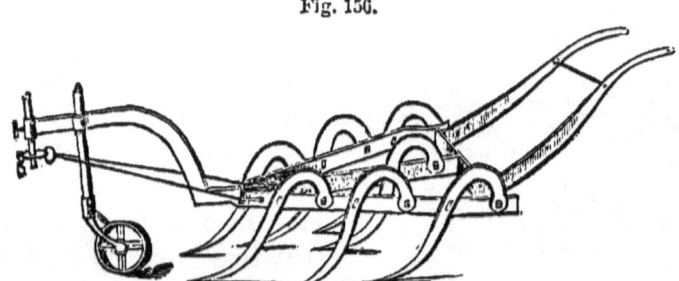

Claw-toothed cultivator for hard ground.

ready moderately mellow. For harder soils, the teeth should be in the form of claws, as shown in fig. 156, their sharp, wedge-form points penetrating and loosening the earth with comparative ease. An efficient cultivator is made by using both kinds of teeth in the same implement, placing the claws forward for breaking the hard earth, and the broader teeth behind for stirring it.

Steel plates, with sharp or "duck-feet" edges screwed at the lower extremities of the teeth, (fig. 157) are useful for paring or cutting the roots of weeds; and formed like

Fig. 157.

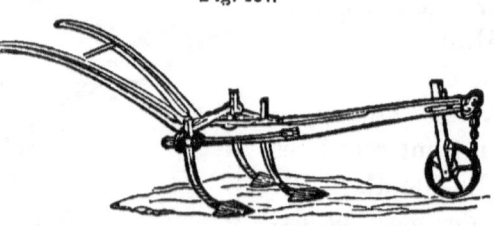

the mould-board of a plow, they are used for throwing the mellow earth toward the row, or, when reversed, from it.

Alden's Thill Cultivator is furnished with fixed thills, extending backwards from the handles. The whole implement thus runs with remarkable steadiness and great efficiency, and the driver, by bearing on the handles, readily increases the depth of the teeth, or by bearing to the right or left, guides it in the row. It is not capable of being expanded and contracted in width.

Garrett's Horse-hoe, an English invention, is a modification of the cultivator, and is used for cultivating carrots and other root-crops in drills, cleaning eight or ten rows at once. It is furnished with sharp, horizontal blades, which run beneath the surface, and shave off and destroy all the weeds within an inch of the rows of young plants. These rows, having been planted by means of a drilling-machine, are straight, and perfectly parallel, and the operator has only to watch *one* row, and guide the blades for that row, the apparatus being so contrived that the blades for the other rows shall run at the same distance from them.

Fig. 158 represents an end view of this implement. It exhibits the apparatus by which the length of the axle is

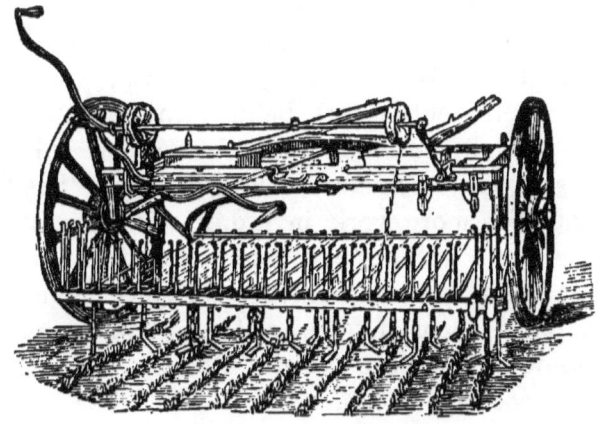

Fig. 158.

Garrett's Horse-hoe—End view.

altered to suit all kinds of planting; by which each hoe is kept independent of the others, so as to suit the inequalities of the ground, and by which they can be set any width, from seven inches to thirty. It shows the oblique angle at which they run—this obliquity being easily altered to any desired degree: this is effected by a movement of the upper handle, represented in the figure. By the lower handle, the whole is accurately guided. It is said that two men, one to lead the horse, and the other to guide the implement, will dress ten acres of root-crops in a single day, and that it has proved eminently a labor-saving machine. It can be used only on smooth land, free from stone.

TWO-HORSE CULTIVATORS

are made to run on two wheels, and the depth of the teeth is regulated by raising or lowering the frame-work that holds them. They have been much used for pulverizing the surface of inverted sod, and fitting it for the reception of seed, but are likely to be superseded for this purpose by Shares' harrow. Modified so as to pass the two spaces between three rows of corn, they are known as double cultivators, and have now come into use for cultivating large fields, and are generally adopted for this purpose at the West. They accomplish twice the work of the single cultivator. They are of two kinds: those called the sulky cultivators, being furnished with a seat on which the driver rides, and the walking cultivators, without seat, the attendant walking behind. The former will accomplish more work in a day, with less fatigue to the driver; the walking cultivators are better suited to rough, or sidling ground, and are cheaper. Many manufacturers make them of different forms, both at the West and in some of the more eastern States. The best sulky cultivators cost about $75.

COMSTOCK'S ROTARY SPADER.

This new machine, which has been used to some extent in the broad fields of the West, forks up the soil by means of a series of revolving teeth. It is drawn by two or four horses, according to its size and the strength of the animals, the driver riding on a seat. Sometimes two machines are attached together, and both are driven by one man. It is used only on land free from sod, such as corn, or other stubble, and is not adapted to land containing stones or rocks.

Its advantages are the following: Greater ease of draught, when compared with the plow, the chief source of friction being the thrusting of the teeth into the soil, while the friction of the plow at the mould-board is usually equal to at least half the weight of the moving sod, added to half the entire weight of both plow and sod, on the sole in the bottom of the furrow, while more force is required to cut with the edge of the share than with the points of the rotary spader. Hence it is found to do twice or three times as much work with the same team as a plow. It does not form a hard crust in the bottom of the furrow, like the plow; and it leaves friable soils pulverized ready for planting, without the use of the harrow.

There are some serious drawbacks to the general introduction of this machine. Its cost exceeds ten times that of a good steel plow, while its complexity renders it more liable to strain or breakage, except in uniform and stoneless soils. It cannot be used in wet seasons, and pulverizes such land only as is previously free from grass. It may, however, prove valuable on extensive farms.

CLOD-CRUSHERS.

In clayey soils, clods are often formed in abundance during the process of cultivation. These become very hard in dry weather, and prevent the proper extension of the fine roots of plants in search of nourishment, and also the intermixture of manure with the soil, without which it has been found that two-thirds, or even three-fourths, of the value of manure is lost to growing crops.

Different modes of pulverizing the clods have been

adopted. The simplest is the "*drag-roller*," represented in fig. 159. It is made of a log, or portion of a hollow tree, into which a common two-horse wagon tongue has been fitted, by which it is dragged over the ground without rolling, grinding to powder, in its progress, every clod over which it passes. The greater the diameter of the log, the less will be the liability of its clogging by gathering the clods before it. It may also be made of a half log, with the round side downward.

Fig. 159.

Clod-crusher.

Fig. 160 represents a similar implement for one horse; this is used for working between the rows of corn in cloddy ground.

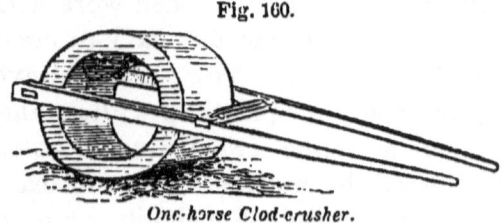

Fig. 160.

One-horse Clod-crusher.

The use of these simple implements, by reducing rough fields to a condition as mellow as ashes, has, in some instances, been the means of doubling the crop. It is necessary that the soil be dry when they are used, to prevent its packing together.

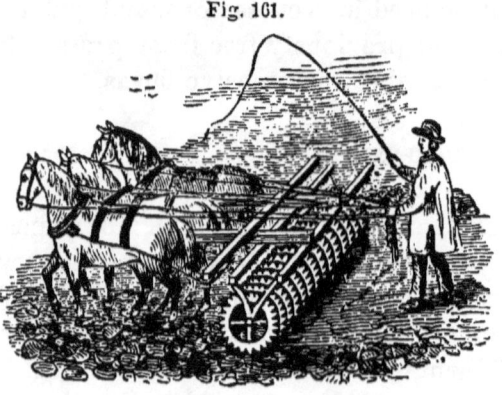

Fig. 161.

Crosskill's Clod-crusher.

Crosskill's Clod-crusher, first used in England, is a more powerful and more costly implement (fig. 161). It

CLOD-CRUSHERS.

consists of about two dozen circular cast-iron disks, placed loosely upon an axle, so as to revolve separately. Their outer circumference is formed into teeth, which crush and grind up the clods as they roll over the surface of the field. Every alternate disk has a larger hole for the axle, which causes it to rise and fall while turning over, and thus prevent the disks from clogging. Fig. 162 represents this implement, as modified and manufactured in this country. It is used only where heavy clay soils prevail.

This clod-crusher can be used only where the ground and the clods have become quite dry. Even then it packs

Fig. 162.

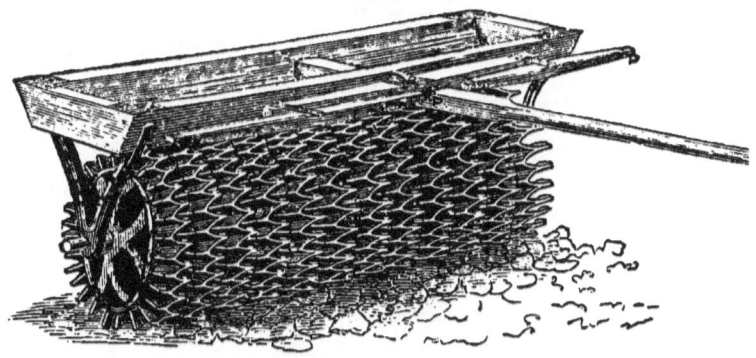

American Clod-crusher.

the soil, and if followed by a harrow, with scarifier teeth, to loosen it again, it would prove an advantage. It is only in certain seasons that it is most successfully employed, or when quite dry weather follows a wet spring. As thorough tile-draining is generally adopted, it becomes less necessary.

The best clod-crushers are sold for about $125.

THE ROLLER.

This implement, now in general use, is employed for pressing in grass seed after sowing, for smoothing the surface of new meadows early in spring, and for other similar purposes. On light soils, it is most valuable, and may be used at nearly all times with safety. Heavy or clay soils will be crusted and injured if rolled while wet. The

Fig. 163.

Field Roller.

roller was formerly made of a single piece, or of a log of wood dressed to a true cylinder; but this scraped the earth when turned to the right or left. A great improvement was made by cutting the single roller into two parts; and a still greater, by employing cast-iron, in several sections, as shown in fig. 163. The cost of cast rollers is about $85 to $100.

CHAPTER XI.

PLANTING AND SOWING-MACHINES.

Sowing-machines, for wheat and other grains, possess great advantages over hand-sowing. All the seed being deposited by them at a nearly uniform depth, and completely covered with earth, it vegetates and grows evenly, and the plants are uniformly strong and vigorous. A less quantity of seed is required, and the crop is heavier.

WHEAT DRILLS.

Several excellent grain drills are now manufactured and sold in this country, having much similarity in external appearance. One of the best and most widely known is made by *Bickford & Huffman*, of Macedon, N. Y. It is represented in the accompanying cut, showing eight dropping tubes. The mode by which the grain is discharged from the hopper down these tubes is exhibited in section in fig. 165; *d* being the interior of the hopper, *bb* a revolving wheel, the projecting rims of which form the bottom of the seed-holder; the axle at *a* causes this wheel to revolve, and the small projections on the interior of the rim carry the seed to *c*, where it drops through an opening in the plate which forms the side of the seed-holder. The rapidity of discharge is perfectly controlled by wheel-work, which causes the axle *a* to revolve slowly or fast at pleasure. The seed-holder is divided into two parts by the wheel *a b*, as shown by cross section in figure 166; one part, *d*, containing wheat, barley, and other medium-sized grains, and the other, *c*, for corn, peas, and the larger seeds. This figure shows the opening in the side-plates, through which the grain is discharged. As these two divisions must be used on separate occasions, the

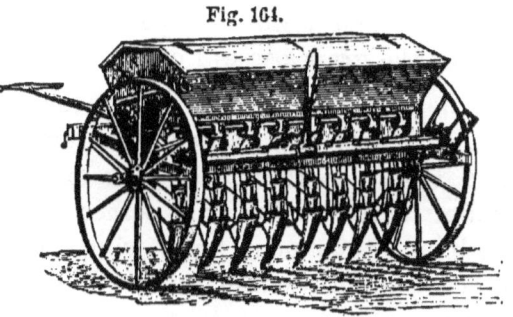

Fig. 164.

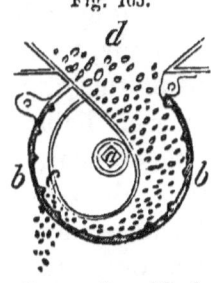

Fig. 165.
Cross-section of Seed-holder.

Fig. 166.
Cross-section of Discharger.

Fig. 167.
Sliding Reversible Bottom of Hopper.

openings between them and the hopper are opened and closed at pleasure by a sliding bottom, with a single movement of the hand. This sliding bottom is shown in fig. 167, and forms hoppers with sloping sides, down which the grain passes freely.

The ends of the tubes, which are shod with steel, are made to pass any desired depth into the mellowed soil, and depositing the seed, it is immediately covered by the falling earth, as the drill passes. This drill is furnished with an attachment for sowing plaster, guano, or any other concentrated manure, and also with a grass-seed sower.

A great improvement has been made in the mode already described, of discharging the seed; formerly, seed-drills generally were furnished with a revolving cylinder, in the surface of which small cavities were made, for carrying off and dropping measured portions of the grain; these often broke or crushed the seed, and were liable to derangement. Others were furnished with circular, revolving brushes, for pressing the seed through holes in the bottom of the hopper; but this contrivance was imperfect, and the brushes were liable to wear out. In the discharging apparatus of the drill just described, the seeds are never crushed, and the whole being substantially made of cast-iron, it may be run a lifetime. The best grain drills are sold for $80 or $90.

SEYMOUR'S BROADCAST SOWER

is an excellent machine for sowing plaster, ashes, guano, salt, or any other concentrated fertilizer, as well as common grain and grass seed. The disagreeable, and even dangerous, as well as heavy and laborious work of sowing these manures by hand renders such a machine desirable on every farm. It is drawn by one horse, sows ten feet wide, and the operator rides in a seat. Seymour's Plaster Sower sows these fertilizers, whether wet or dry. These machines are sold at about $70.

CORN PLANTERS.

Among the best one-horse corn planters, which make one drill at a time, are Emery's, Harrington's, and Billings'. The last-named is represented in the annexed cut. It drops in hills, eleven inches apart in the row, or, if desired, twenty-two inches, the perforations in the slides regulating the number of grains. It is so constructed as to drop any desired amount of plaster, guano, or other concentrated manure, without coming in contact with the seed. This, and other one-horse drills, are well adapted to planting fields of considerable size, for cultivating in rows but one way. On a larger scale, two-horse drills are employed. Wheat drills are often used for this purpose, employing only two of the tubes. Another class of corn planters, for planting in hills, the rows running both ways, consist of hollow tubes, which contain the seed, and which, by striking or pressing on the soil, drop and cover a hill at one stroke.

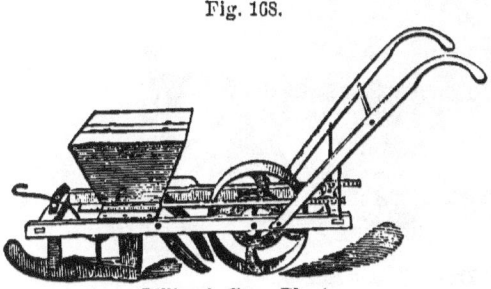

Fig. 168.

Billings' Corn Planter.

TRUE'S POTATO PLANTER.

For field culture, this implement has proved an important saver of hand labor. It is drawn by one horse, and cuts, drops, and covers the potatoes at one operation. It is usually employed on ground which has been plowed and harrowed only, the driver forming the drills by the eye, as the planting proceeds. Straighter rows may be made by first marking the land with a good corn-marker, and then employing a small boy to ride, directing him to keep the horse on the line. The driver has then only to

Fig. 100.

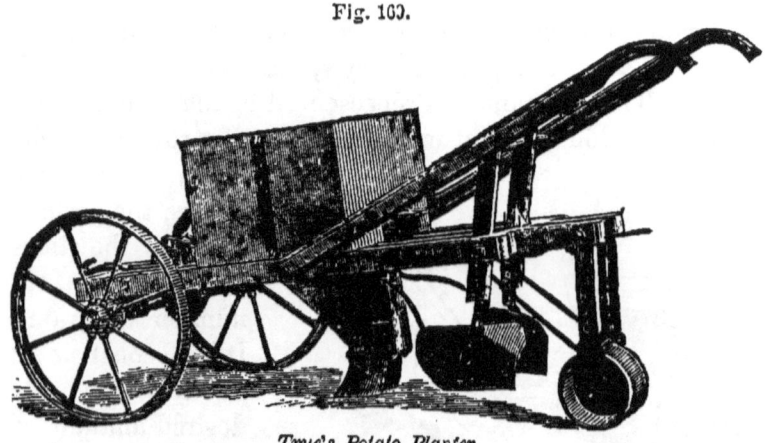

True's Potato Planter.

watch the working of the machine before him. If the ground is rough, or rather dry, it is better to furrow the land previously with a single horse, running the planter in these furrows.

For using this machine successfully, the seed potatoes must be previously assorted, so that those of nearly equal size may be used at a time. It is common to assort them into two sizes, which may be done in winter, or on rainy days. Each potato passes the throat of the hopper singly; and if one in a bushel happens to be too large, it will choke the opening. After passing the hopper, each potato is sliced into pieces of the desired size, which then, one by one, drop down the hollow coulter, and are buried. The throat of the hopper is readily contracted or expanded, and adapted to any assorted size of seed. One man, with a horse, will plant several acres in a day, and if the ground be in good order, with nearly or quite as much accuracy as by hand, and with more uniformity of depth.

HAND DRILLS, OR SEED SOWERS.

These are great savers of labor for sowing the seeds of ruta-bagas, carrots, field beets, and other farm root crops, besides peas and beans. One of the best in use is Harrington's, represented by fig. 170, and made by F. F. Holbrook & Co., Boston. The side chains mark the rows, and it makes its own drill, drops, and covers the seed with accuracy, at one operation. It is readily changed to the hand cultivator, by removing the dropper, and attaching the cultivator teeth, shown in fig. 171. It then becomes a convenient implement for running between the rows, in small fields.

Fig. 170.

Harrington's Hand Seed Sower.

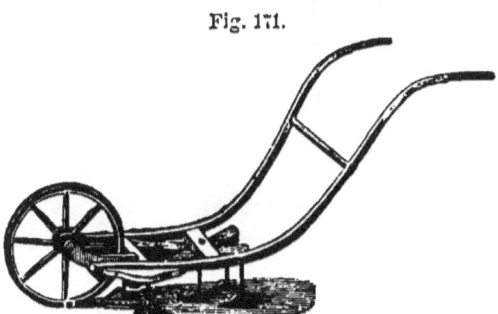

Fig. 171.

Harrington's Hand Cultivator.

CHAPTER XII.

MACHINES FOR HAYING AND HARVESTING.

MOWING AND REAPING MACHINES.

The cutting part of the mowers and reapers made at the present day consists of a serrated blade, as shown by

Fig. 172.

Knives.

fig. 172, which passes through narrow slits in each of the fingers, shown in fig. 173, forming, when thus united, the cutting apparatus, as exhibited in the annexed figure, of *Wood's Mowing-machine* (figure 174). When the machine is used, the motion of the wheels on which it runs is multiplied by means of the cog-wheels, imparting quick vibrations, endwise, to this blade, shearing off the grass smoothly as it advances through the meadow, like a large number of scissors in exceedingly rapid motion.

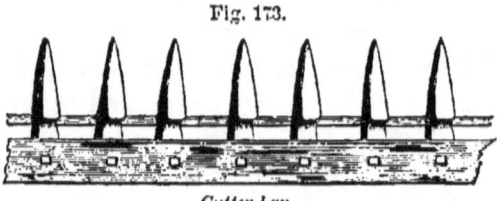

Fig. 173.

Cutter-bar.

Fig. 174.

Wood's Mower.

The finger-bar, the most important part, now adopted

in all mowing and reaping machines, was invented by Henry Ogle, of Alnwick, England, in 1822, and his machine was put in successful operation, after much experimenting, by T. & J. Brown, of that place. But so strong was the prejudice of the working people against labor-saving machinery, that they threatened to kill the manufacturers if they persevered; and the enterprise for a time was given up.*

The limits of this work permit only a brief notice of some of the chief points in mowers and reapers; and a few machines are referred to, out of a large number of kinds, which are made in the different States, and which have proved themselves worthy of the confidence of farmers.

Fig. 175.

The Kirby Machine as a Mower.

Fig. 176.

Buckeye Mower with Folded Bar.

The operation of mowing is shown in fig. 175, which represents the Kirby mower, one of the best single-wheel machines, cutting a swath five feet wide, as fast as the horses advance.

Various contrivances are adopted for lifting or folding the cutter-bar when the machine is not in operation, or in passing from one field

---

* Woodcroft.

to another. A neat and convenient form is used in the Buckeye Mower, represented in the accompanying cut, (fig. 176) where the bar is folded over in front of the driver's feet.

In the mowing-machine, the cutting apparatus is narrow, causing the newly cut grass to fall evenly behind it, covering the whole surface of the ground. The reaping-machine is similar in construction, with the addition of a platform for holding the grain as it falls, as shown in the annexed figure of the Kirby machine, changed to a reaper (fig. 177).

Fig. 177.

Kirby Reaper, with Hand Rake.

This figure represents the reel, which is attached to, and is worked by the machine, causing the grain, as it is cut, to drop smoothly upon the platform. When a sufficient quantity has collected there, it is swept off by the hand rake, and is afterwards bound in a sheaf. The annexed cut exhibits the Cayuga Chief, (an excellent two-wheeled machine) as a reaper, in which the operation of hand-raking is distinctly represented.

Fig. 178.

Cayuga Chief—Combined Mower and Reaper.

SELF-RAKING REAPERS.

Mowing-machines need but one man for their management, who merely drives the horses that draw it.

Reapers, as usually made, require another man besides the driver, to rake off the bunches of cut grain, which is severe labor. Various *self-raking* contrivances have been used to obviate this labor, several of which have been made to do excellent work, and are coming into general use.

One of the first successful self-raking attachments to the reaper was that used by Seymour & Morgan, of Brockport, N. Y. It was one of the kind which sweeps across the platform, in the arc of a circle, delivering the gavel at the side of the machine. The ordinary reel is used with this class of rakes. An objection to them is, that the grain is seized for throwing off at a point behind the cutters. Owen Dorsey introduced an improvement in the form of what are termed reel-rakes, which strike the grain forward of the cutters. A series of sweeps or beaters were employed, combined with one or more rakes, the gavel being delivered from the platform at each circuit of the rake. At first, the horizontal motion of these arms prevented the driver from riding on the machine. An improvement was effected, so that the arms and rakes, after passing the platform, were made to rise to a nearly vertical position, thus passing the driver freely. The

Fig. 179.

The Kirby Self-raker.

accompanying engraving, (fig. 179) representing the self-raker used on the Kirby machine, shows the position of the arms when in motion—one of them serving as a rake at each revolution. There are several modifications of this class of rakes, made by different inventors. *Marsh's* machine consists of beaters and rakes combined, and delivers one or more gavels at each revolution, according to the number of rakes used at a time. *Johnson's* rake is furnished with rake-heads for each of the arms, which are so arranged as to dip low into the grain forward of the cutters, and afterwards to rise in passing over the platform. To discharge the grain, the driver uses a latch-cord and lever, so that the path in which the rake travels is changed by opening a switch or gate, permitting one of the rakes to pass low enough to sweep the platform. The Cayuga Chief, Buckeye, Hubbard, and other reapers, use this self-raker. The *Kirby* machine employs a self-raking attachment of its own, already represented in fig. 179. Two or three of the arms, or beaters, at the option of the driver, bring the grain on the platform; the other one or two carry the rake-head. The driver may throw off a gavel, or two gavels, at each revolution; or the rake may be made to run continuously, at regular intervals, without attention on the part of the driver. The arms, or rakes, are so made as to be adjustable to the height of the grain.

The Dropper is a simple contrivance, (represented in the annexed engraving) consisting of a light platform, which holds the grain until the gavel is large enough, when it suddenly drops and discharges it. It is much used at the West, and, although hardly so perfect as some self-rakers, is preferred by many farmers, the gavels being delivered behind the machine, and thus keeping the binders up to their work, in clearing the way for the next passage of the reaper.

Fig. 180.

Cayuga Chief with Dropper.

BINDERS.

Several machines for binding grain have been invented, possessing considerable merit, but so far they do not appear to be adapted to general introduction.

Marsh's Harvester, much used at the West, is so constructed, that two men may readily bind as fast as the harvester does its work. The binders stand on a small platform, furnished with a guard or rail, and the grain, as fast as it is cut, is carried up by an endless apron to a platform, where each man alternately makes his band, and receives and binds his sheaf. As they expend no time in stooping, or in passing from gavel to gavel, they are enabled to work with ease and rapidity. The weight is only that of one man more than on a hand-raker.

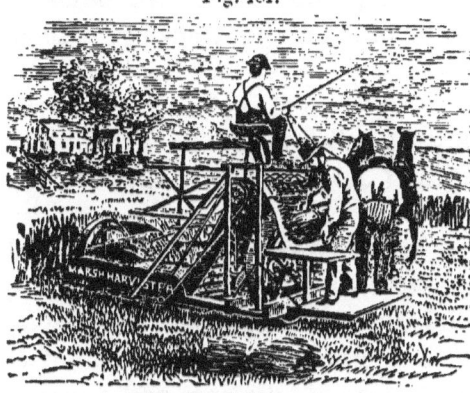

Fig. 131. The Marsh Harvester.

Headers are reaping-machines employed for cutting the heads of wheat with a small portion of the straw, leaving most of the straw standing. They are usually driven by four horses, and are thrust forward ahead of the team. A two-horse wagon, in addition, is driven along side, to receive from an endless apron the heads, as they are cut by the reaper. They are only used on the extensive fields of the West, and a difference of opinion prevails as to their general value.

DURABILITY AND SELECTION.

Mowing and reaping-machines, being complex, or made up of many parts, would soon be broken and destroyed, if the resistance they meet with were irregular and full of obstructions, like those which the plow encounters. Standing grain and grass present a soft and uniform resistance, and hence, well-made machines will last several years without much repair. The Report of the Auburn trial of mowers and reapers gives five years as the average "lifetime" of these machines. Much will depend on the amount of work performed in a season; an extensive farmer states, that he usually cuts about five hundred acres with each machine before it needs renewing. Much, also, depends on the care which the machines receive; such as keeping them always well sheltered from the weather, and thoroughly cleaning every part, and carefully wiping the journals and bearings before they are laid aside for the season.

In selecting mowers and reapers, there are several points which the purchaser should carefully observe; as, for example—1. Simplicity of construction. 2. Use of best material for knives and other parts used in manufacture. 3. Finish and perfection of gearing and running parts. 4. Durability, as proved by use. 5. Ease of draught. 6. Freedom from side draught. 7. Quality of work. 8. Ease of management. 9. Convenience and safety of driver. 10. Adaptation to uneven surfaces. A part of these points can be fully determined only by thorough trial; and it is always safest to purchase of those manufacturers whose machines have been long enough in general use to establish their character in these respects. Fortunately, there are many in different parts of the country, who have secured a good reputation, from whom machines, or parts for repairs, may be obtained without sending long distances. The report of the Auburn trial, in 1866, states, that out of twenty different mowing-machines, which were tried on a rough meadow, every one, with two exceptions, "did good work, which would be acceptable to any farmer; and the appearance of the whole meadow, after it had been raked over, was vastly better than the average hand mowing of the best farmers in the State." Since that trial, a continued improvement in manufacture has been taking place, and the machines are becoming more perfect.

The price of a good two-horse mowing-machine is about $120; and of a combined mower and reaper, about $170.

HAY TEDDING MACHINES.

Machines for stirring up and turning the drying hay have long since been known and used in England, and a few were introduced into use in this country. But as they were heavy and cumbersome, they never came into common use. A few years since, Bullard's Hay Tedder was invented, and has been widely used. It scatters and turns the hay with great rapidity, and consists of several forks, held nearly upright, but worked by a compound crank, so as to scatter the hay in the rear of the machine.

Fig. 182.

Bullard's Hay Tedder.

The close resemblance of the movement of these forks to the energetic scratching of a hen presents a ludicrous appearance to one who sees it for the first time. The use of the tedder is found greatly to hasten the drying process, especially on heavy meadows, and to enable the farmer to secure his hay in so short a time as frequently to avoid damaging storms.

A new machine, remarkable for its simplicity and perfection of working, is the *American Hay Tedder*, made by the Ames Plow Company, of Boston. It is represented in the accompanying cut. It is furnished with sixteen forks, attached to a light reel in such a manner that they revolve rapidly, with a rotary, continuous, and uniform motion. It never clogs, may be easily backed, and readily passes over ordinary obstructions, without any attention on the part of the driver.

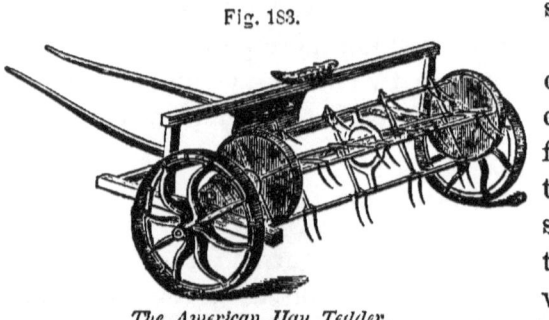

Fig. 183.

The American Hay Tedder.

Hay tedders should be used on the meadow about three times a day, which will enable the farmer to cut his crop in the morning, and draw it in the same day; giving him, also, more uniformly dried, and better hay.

The price of hay tedders varies from $75 to $100.

HORSE HAY-RAKES.

The simplest and original form of the horse-rake is represented in fig. 184. It was made of a piece of strong scantling, three inches square, tapering slightly toward the ends, for the purpose of combining strength with lightness, and in which were set horizontally about fifteen teeth, twenty-two inches long, and an inch by an inch and three-fourths at the place of insertion, tapering on the under side, with a slight upward turn at the points, to prevent running into the ground. The two outer teeth were cut off to about one-third their first length, and draught-ropes attached. If these pieces were too short, the teeth were hard to guide; if too long, the rake was unloaded with difficulty. Handles served to guide

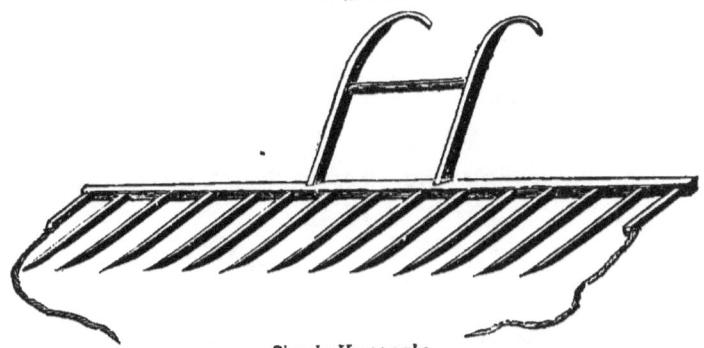

Fig. 184.

Simple Horse-rake.

the teeth, to lift the rake from the ground in avoiding obstructions, and to empty the accumulated hay.

In using this rake, the teeth were run flat upon the ground, passing under, and collecting the hay. When full, the horse was stopped, the handles thrown forward, the rake emptied and lifted over the windrow thus formed. The windrows, as in other horse-rakes, were made at right angles to the path of the rake, each load being deposited opposite the last heap formed, in previously crossing the meadow. A few hours' practice enabled any one to use this rake without difficulty; the only skill required was to keep the teeth under the hay, and above the ground.

In addition to raking, this implement was employed for sweeping the hay from the windrow, and drawing it to the stack. It was also useful for cleaning up the scattered hay from the meadow, at the close of the work; for raking grain-stubble, and for pulling and gathering peas. If made of the toughest wood, and with the proper taper in the main parts for lightness and strength, according to the principles already pointed out in a previous chapter, it was easily lifted, and its use not attended with severe labor.

This simple horse-rake has nearly gone out of use, and yet, on account of its simplicity and cheapness, it is worthy of being retained on small farms, and especially on meadows with uneven surfaces. The cost need not be more than three or four dollars. From twelve to fifteen acres could be raked with it in a day.

The *Revolving Horse-rake* (fig. 185) was next generally adopted, possessing the great advantage of unloading

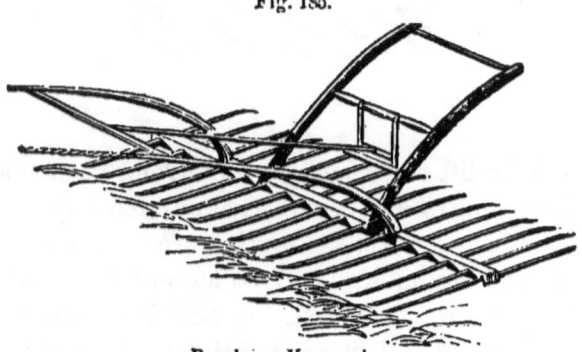

Fig. 185.

Revolving Horse-rake.

without lifting the rake or stopping the horse. It has a double row of teeth, pointing each way, which are brought alternately into use as the rake makes a semi-revolution at each forming windrow, in its onward progress. They are kept flat upon the ground by the pressure of the square frame on their points, beneath the handles; but as soon as a load of hay has collected, the handles are slightly raised, throwing this frame backwards, off the points, and raising them enough for the forward row to catch the earth. The continued motion of the horse causes the teeth to rise and revolve, throwing the backward teeth foremost, over the windrow. In this way, each set of teeth is alternately brought into operation. The cost of this rake is from $7 to $10, and twenty acres or more could be raked with it in a day.

A further improvement has been made in the revolving rake, by attaching it to a sulky, on which the operator rides, enabling him to do a larger amount of work with less fatigue. There are several modifications, some of which place the rake in front of the sulky wheels, and others, in the rear. One of the best and most widely used is "Warner's Sulky Revolver," manufactured by Blymyer, Day & Co., of Mansfield, Ohio, and by others. It is represented in the annexed cuts, fig. 186 showing it in the operation of raking, and fig. 187, the same machine, with the rake thrown upon the wheels, for driving from field to field. The head is the same as the common revolving rake-head—the teeth being tipped with malleable iron. The rake is operated by means of a lever, attached to a journal at the centre of the rake-head. By means of cams, stop, and spring, the lever and head are entirely at the will of the operator. A slight pressure, equal to seven or eight pounds, on a lever, causes the rake to revolve; and it is also readily elevated for backing, or for passing obstructions.

Fig. 186.

Fig. 187.

An important advantage of this rake is, its gathering the hay free from gravel and earth; also its cheapness recommends it—the price being about $35.

SPRING-TOOTH RAKE.

The original form of the spring-tooth rake is shown in fig. 188. The teeth were made of stiff, elastic wire, on the *points* of which the rake ran, and not on the flat sides, as in those already described. They bent in passing an obstruction, and sprung back again to their place. This rake was unloaded by simply lifting the handles, which was easily done, the rake being light, and about one-half the weight being sustained by the horse.

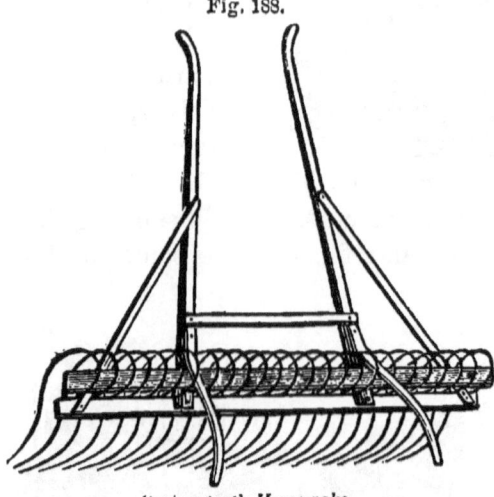

Fig. 188.

Spring-tooth Horse-rake.

All the spring-tooth rakes made and used at the present time are attached to wheels, and a seat is furnished for the driver. There are many patented modifications, some possessing advantages of greater simplicity, or ease of management; but all appear to be good and efficient rakes, enabling the operator to gather about twenty-five acres in a day.

Among the best of the spring-tooth rakes is that of Hollingsworth, made by Wanzer & Cromwell, of Chicago, and represented in the accompanying engraving. Each tooth is separate, and may be readily replaced. As soon as the rake is loaded with hay, the driver, by touching the lever before him, drops it at the line of the windrow. The cost is about $45,

THE HAY-SWEEP.

Fig. 189.

Hollingsworth's Spring-tooth Rake.

THE HAY-SWEEP.

Where the hay is secured in stacks, or in hay-barns situated contiguous to the meadow, the use of the hay-sweep, in connection with the horse-fork, would probably enable two or three men, and two boys, with three horses, to draw and pack away *thirty tons a day, or more.* The hay-sweep, invented many years ago by W. R. Smith, of Macedon, N. Y., is but little known. The accompanying figures (190 and 191) exhibit its construction and use. It is essentially a large, stout, coarse rake, with teeth pro-

Fig. 190.

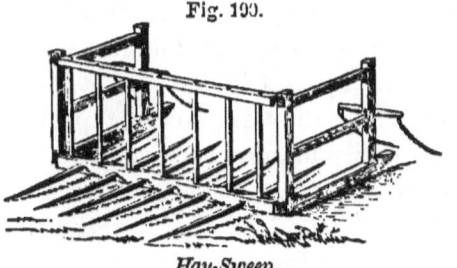

Hay-Sweep.

jecting both ways, like those of a common revolver; a horse is attached to each end, and a boy rides each horse. A horse passes along each side of the windrow, and the two

Fig. 191.

Hay-sweep in Operation.

thus draw this rake after them, scooping up the hay as they go. When 500 pounds or more are collected, they draw it at once to the stack, or barn, and the horses turning about at each end, causing the gates to make half a circle, draw the teeth backward from the heap of hay, and go empty for another load—the teeth on opposite sides being thus used alternately. To pitch easily, the back of each load must be left so as to be pitched first.

The dimensions should be about as follows:—Main scantling, below, 4 by 5 inches, 10 feet long; the one above it, same length, 3 by 4 inches; these are three feet apart, connected by seven upright bars, 1 by 2 inches, and 3 feet long. The teeth are flat, 1½ by 4 inches, 5 feet long, or projecting 2½ feet each way; they are made tapering to the ends, so as to run easily under the windrow. A gate, swinging half way round on very stout hinges, is hung to each end of this rake, and to these gates the horses are attached. Each gate consists of two pieces of scantling, 3 inches square, and 3 feet long, united by two bars of wood, 1 by 2 inches, and a third, at the bottom, 3 inches square, and tapering upwards, like a sled runner;

these runners project a few inches beyond the gate. The whiffle-trees are fastened a little above the middle of the gate, and should be raised or lowered so as to be exactly adjusted. This machine may be made for $6 or $7.

In using, not a moment is lost in loading or unloading. No person is needed in attendance, except the two small boys that ride the horses. If the horses walk three miles an hour, and travel a quarter of a mile for each load, they will draw 12 loads, or three tons an hour, or thirty tons in ten hours, leaving the men wholly occupied in raising the hay from the ground, by means of another horse, with the pitchfork.

It will be obvious, that this rapid mode of securing hay will enable the farmer to elude showers and storms, which might otherwise prove a great damage.

HORSE HAY-FORKS.

Every farmer who has ever pitched off from a wagon in one day ten or twelve tons of hay is aware that no labor on the farm can be more fatiguing. The horse-fork, in its various forms, which, to a considerable extent, has been brought into use, has afforded great relief, severe labor being not only avoided, but much greater expedition attained. The effective force of a horse is, at least, five times as great as that of a stout man; and if half an hour is usually required for him to unload a ton of hay, then only six minutes would be necessary to accomplish the same result with horse-power. Actual experiment very nearly accords with this estimate.

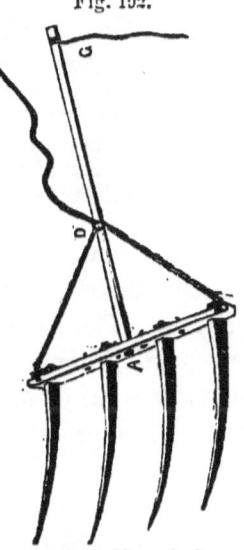

Fig. 192.

Original Horse-fork.

A simple form of the horse pitchfork was described in

the Albany Cultivator, in 1848, from which a subscriber in Bradford County, Pa., made the first used in that region. Some years later, he stated that there were at least two hundred in use. The preceding figure represents this simple and original fork. *A* is the head, twenty-eight inches long, and two and a half inches square, made of strong wood. *A G* is the handle, five and a half feet long, mortised into the head, with an iron clasp or band of hoop iron fitting over the head, and extending six inches up the handle, secured by rivets. The prongs of the fork are made of good steel, one-half an inch wide at the head, twenty inches long, and eight inches apart, with nuts to screw them up tight. Rivets are placed on each side of the middle ones, to prevent the head from splitting. The rope is attached to staples at the ends of the head. The single rope *D* extends over a tackle-block, attached to a rafter at the peak of the barn, about two feet within the edge of the bay. The rope then passes down to the bottom of the door-post, under another tackle-block, and to the outside of the barn, where the working horse is attached to it. A small rope or cord *G* is attached to the end of the handle, by which it is kept level, as it ascends over the mow. The cord is then slackened, and

Fig. 193.

Pitching Hay through a Window with Horse-power.

the hay tilts the fork, discharging its load. The horse is then backed up, ready for another fork load, the only labor of the workman being to drive the fork into the hay and keep the cord steady. An important advantage is gained, besides the saving of time; for the man on the load, being relieved from the severe labor of pitching, is fresh and vigorous for throwing on another load in the field.

The length of the handle made it difficult to use this fork under low roofs, and an improvement was made by *Gladding*, by which the head of the rake only was tilted, leaving the handle in its horizontal position. A hinge-joint is placed at the connection of the head and handle, so that, at any moment, by a jerk on the cord which passes up a bore in the handle, the fork is dropped, as shown in fig. 194, and its load deposited. This may be done instantaneously, at the moment it happens to be swung to the most favorable spot.

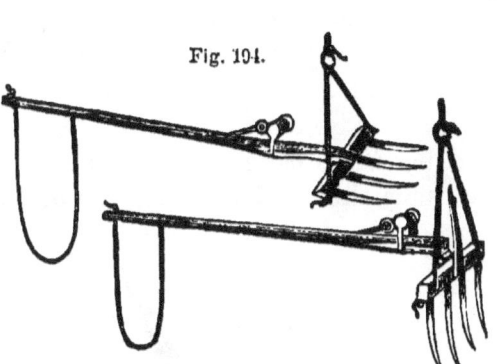

Fig. 194.

Gladding's Hay-fork.

Its weight causes the head to fly back of its own accord, and resume its former position, ready for another forkful. The rope suspending the fork should be fastened to the highest portion of one of the rafters, over the mow, and a smooth board should be placed, vertically, against the face of the mow, for the hay to slide on as it ascends. By attaching this rope in front of, and within a window, the hay is carried with ease into the window, and thus lofts over sheds, carriage-houses, etc., where the old horse-fork could not be used, are filled by the use of *Gladding's* improvement. This is one of the best forks, adapted to all kinds of pitching, and has unloaded a ton of hay in about three minutes; and over a beam twenty-two feet high, under a low rafter, in about nine minutes.

In using horse forks, as already stated, their operation is much facilitated by providing a board slide, to be placed vertically against the face of the mow, or bay, on which the hay moves upward. In pitching into a window, the bottom of this board slide should be placed out a few feet from the building, and the top should rest on the base of the window. When convenient, the back end of the wagon load should be placed towards the window. There is no limit to the height at which the pitching may be easily performed—giving the use of the horse-fork a great advantage over hand pitching; and barns, with high posts, may be built for the storage of hay.

Other forms have been adopted for pitching under roofs, by using shorter handles. One of the best is *Palmer's*

Fig. 195.   Fig. 196.

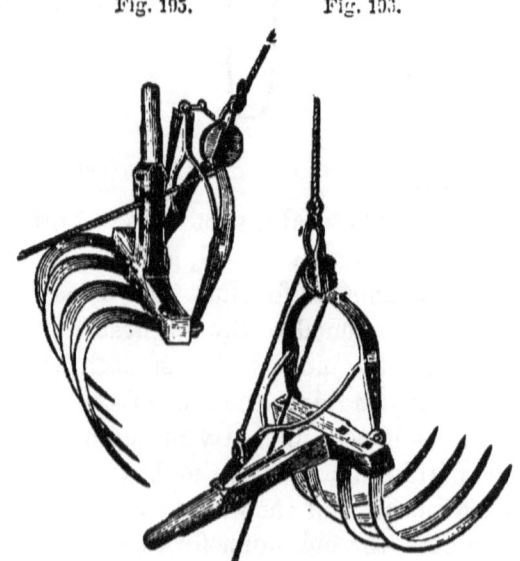

Palmer's Fork.

Fork, made by Wheeler & Co., Albany, and Palmer & Co.,

DOUBLE FORKS. 177

Chicago, which is represented in the accompanying figures, the right-hand one showing its position when ascending, loaded with hay; the left-hand, with the knee-joint brace contracted, by jerking the cord for emptying the load. Still another, known as *Myers' Elevator*, is shown in fig. 197, in its position when lifting the hay, and fig. 198, when dropping it. The head is iron, and it is a strong and simple fork.

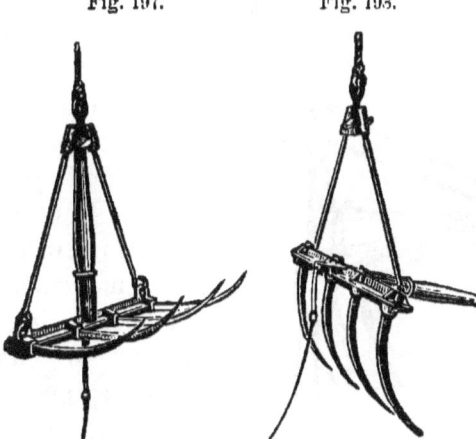

Fig. 197. Fig. 198.

Myers' Hay Elevator.

DOUBLE FORKS.

The double forks clasp the load of hay like the claws of a bird. This class of forks may be used for pitching over a beam, without a board facing. They are better adapted to pitching short straw, especially those which like Raymond's, have several teeth; but more time is required for thrusting in the two forks than one.

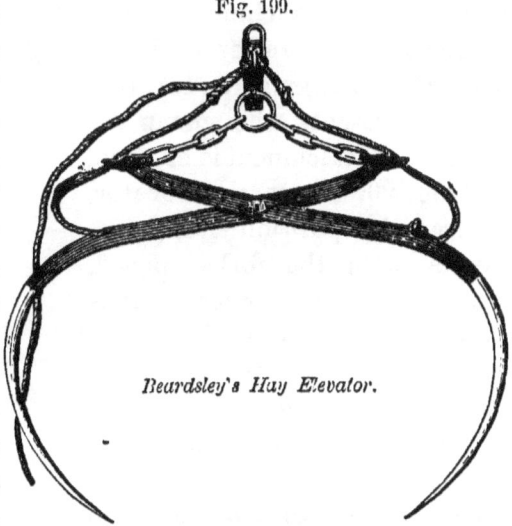

Fig. 199.

Beardsley's Hay Elevator.

One of the simplest is *Beardsley's Hay Elevator*, (fig. 199)

8*

which sufficiently explains itself.

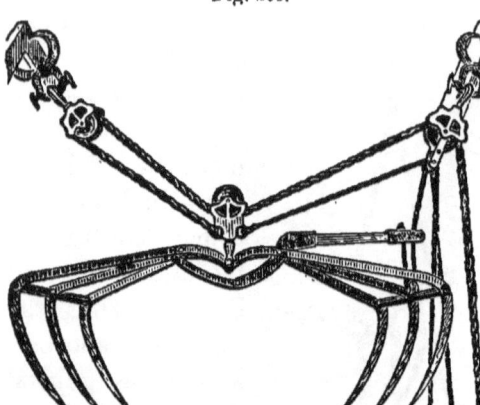

Raymond's Fork.

The "Little Giant" Fork resembles the claws of a bird, and has a fluted, tubular, cast head, the single grasping-tooth being double-jointed, and permitting it to enter the grain freely. On the movement of the horse, it is brought to its place, grasping its load firmly. *Raymond's Elevator*, made by J. H. Chapman, Clayville, N. Y., consists of two three-pronged forks, connected together by a hinge, (fig. 200) and is one of the best double forks. Connected with this fork is a ready contrivance for attaching it, in a moment, to any rafter or beam. The accompanying figure (fig. 201) represents the clamp by which this attachment is effected, and fig. 200 shows the elevator, secured in position from two points, with the forks opened, when dropping their load. It is raised and lowered by the double ropes passing over the two fixed pulleys, and the one on the elevator—the horse moving twice as fast as the load is raised.

Grappling Irons and Hoisting Tackle.

Thus attached to two beams, the load may be run horizontally, as well as raised vertically, as more fully explained under the head of *Stacking*. By the single fastening, (fig. 201) the fork is only raised vertically.

HARPOON FORKS.

For pitching hay exclusively, or any material which hangs well together, the harpoon forks do their work more rapidly than any other, but they are not adapted to

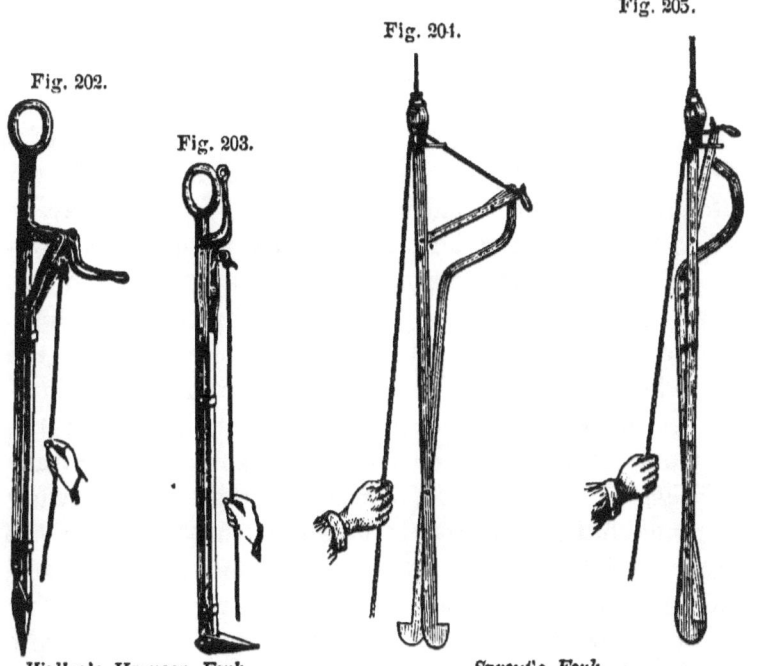

Fig. 202. Fig. 203. Fig. 204. Fig. 205.

Walker's Harpoon Fork.   Sprout's Fork.

short straw. *Walker's* harpoon, made by Wheeler, Melick & Co., Albany, is a straight bar of metal, appearing almost as simple as a crow-bar, (fig. 202. Its point is driven into the hay as far as desired, when a movement at the handle is made, which turns up the point at right angles, (fig. 203,) enabling it to lift a large quantity of hay. A modification has spurs, which are thrown out on opposite sides. The combined fork and knife invented by Kniffen & Harrington, of Worcester, Mass., is an excellent hay-knife, when folded, as in fig. 205, and an efficient elevator, when opened, as in fig. 204. It is well adapted to the use of farmers who have nothing but hay to pitch, and plenty of room for the elevator to swing in. At the Auburn trial, this fork discharged a load of hay weighing twenty-three hundred pounds, over a beam, in two minutes.

The prices of horse-forks, of different kinds, vary from $10 to $20.

HAY CARRIERS.

An inconvenience results from the fixed position of a hay-fork, preventing the hay from being distributed over different parts of a broad bay, except so far as it may be swung to the right or left, and the load dropped at a signal. Several hands are sometimes required to spread this hay evenly, as it is rapidly discharged by the horse-fork. Another disadvantage is, the required narrowness of the bay, which cannot well be more than twenty or twenty-five feet wide. These objections are obviated, and the hay carried fifty or a hundred feet horizontally, by means of *Hick's Elevator and Carrier*, of which the following clear and full description is given in the Report of the Auburn Trial of Implements:—" It consists of a track, made of 2 by 5-inch plank, fastened to the rafters a few inches below the ridge of the barn by 1½-inch square strips and twelve-penny nails. Upon this track runs a car; a rope passes through it, and through a catch pulley attached to a horse hay-fork, then back to the car; the other end passes back to the end of the barn, and returns through pulley wheels to the barn floor, to which end a horse is attached.

By a peculiar arrangement of the car, it is held in position on the track, over the load to be unloaded, until a forkful of hay is elevated to it, when it is liberated from its position, and the fork made fast to the car in one operation, then it moves off on the track very easily, and any distance you may choose to have it carried; the operator, by pulling a cord, trips the fork, and the horse, turning around, walks or trots back to the place of starting; the car is pulled back to its position by the trip cord, when the fork descends for another load.

The fork comes back so easily and quickly that the horse can be kept in motion continually, elevating from 300 to 400 pounds of hay, and carrying it forty to fifty feet in a horizontal direction, and returning for another load in less than a minute.

Its advantages over the old mode are:

1st.—The hay can be carried into the second, third, and fourth bays from the wagon, as easily as into the first, thus saving a large amount of labor in the mows.

2d.—The hay is elevated perpendicularly from the load, thus obviating the friction caused by dragging the forkful of hay over and against the beam; also the danger of tripping or breaking the fork as it is drawn over the beam.

3d.—The car and fork return so easily, the fork dropping in the middle of the load, ready to be thrust into the hay immediately; whereas, in the old method, it is very hard work to get the fork back, if the hay has been carried any distance.

4th.—The horse turns around, and walks or trots back to the place of starting, instead of backing, thus saving much labor to both horse and driver.

5th.—The hay need be elevated only high enough to clear the highest beam, when it can be carried horizontally, until the mows are more than half full, when, by shortening a rope, the fork can be made to pass along only sixteen inches below the very peak of the barn.

6th.—It requires but very little force to carry the hay horizontally, whereas, by the old methods, it requires more force to carry it horizontally than to elevate it.

7th.—By extending the track four feet beyond the end of the building, hay can be elevated and carried into long, low hovels, or cow barns, when no other arrangement would work at all.

The car is small, and the track light and simple; a weight has been lifted of 1,080 pounds by it at one time, with a pair of mules."

By using a strong car, it may be employed for unloading coal from a boat.

BUILDING STACKS.—Three long poles may be used for this purpose, securely chained at the top, and spread in the form of a tripod. The one to which the lower pulley is attached should be set firmly into the ground, to prevent displacement by the outward draught. Holes are bored into the poles at convenient distances, and cross pieces secured to them, for holding the board slide, and permitting it to be gradually raised, as the stack goes up. The hay may be pitched from the ground as well as from a load, without inconvenience, to any height.

Fig. 206.

Mode of Coupling the Poles.

Instead of chaining the poles together, they may be firmly secured by using two stout clevises, the bolts of which are passed through auger holes, near the upper ends of the poles, (fig. 206).

PALMER'S HAY STACKER, represented in fig. 207, has been much used at the West, where large quantities of hay are deposited out of doors. It first elevates the hay, and then swings it around over the stack, dropping it where desired. It does not drag the hay against the side of the stack, requires no staking down to prevent tipping, and is easily drawn on the sills as runners, to any part of the farm. The horizontal motion of the crane is effected as follows:—Two ropes are attached to the whiffletree, one, a strong one, to elevate the hay, running on the pulleys at B, C, and D; and the other, a smaller one, passing the swivel pulley at A, on the end of the lever B, extending from the foot of the upright shaft. This cord then passes up and over a pulley above the weight E. The weight is about four pounds, and is attached to the end of the smaller cord. At the same time that the horse, in drawing, elevates the fork with its load of hay, the weight E is raised until it strikes the pulley, when the power of the horse becomes applied to the end of the lever B, causing it to revolve, and swing the hay over the stack. As the horse backs, the weight drops again to the ground, taking up the slack rope from under the horse's feet, and the weight of the fork causes the arm of the derrick to revolve back over the load. The intended height for raising the hay, before swinging, is regulated by lengthening or shortening the smaller cord, as the arm will not revolve until the weight strikes the pulley under the head block. T. G. & M. W. Palmer, of Chicago, own this invention, and furnish the smaller parts of the machine, the heavier being easily made on the farms where intended to be used.

Fig. 207.

Palmer's Hay Stacker.

Fig. 208 shows the manner in which Raymond's Elevator is mounted for stack building. These poles need not be so heavy as when three poles alone are used. They are kept from being drawn over towards each other in elevating heavy loads, by lashing the lower end of each outer pole to a strong stake, driven into the ground obliquely, by first making a hole with a crow-bar. It is convenient to place the two pole tripods sufficiently distant from each other to give room for the stack, or rick,

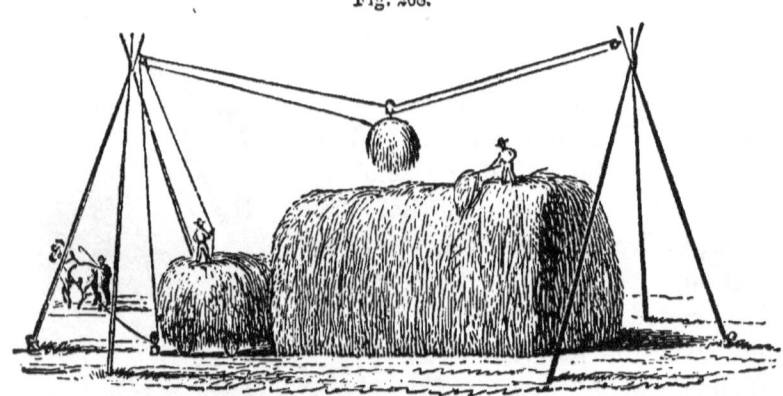

Fig. 208.

Fork on Poles for Building Stacks.

and to allow the wagon to pass within them. The elevator first lifts its load, and then carries it along the rope, till the man on the load drops it by a jerk of the cord. This apparatus is made by J. H. Chapman, of Clayville, N. Y.

HAY PRESSES.

Among the best Hay Presses in the country is the one manufactured by L. & P. K. Dederick, Albany, and represented in the annexed engraving. It is worked by one or two horses, operating with great force by means of the arms on each side, which are connected with toggle-joint levers, explained in a former part of this work. The hay is thrown in from the upper platform, and when reduced to compact bales, by means of the powerful force which this press gives, is taken out at the lower. In order to prevent the necessity of the horses running back at

DEDERICK'S HAY PRESS.

Fig. 209.

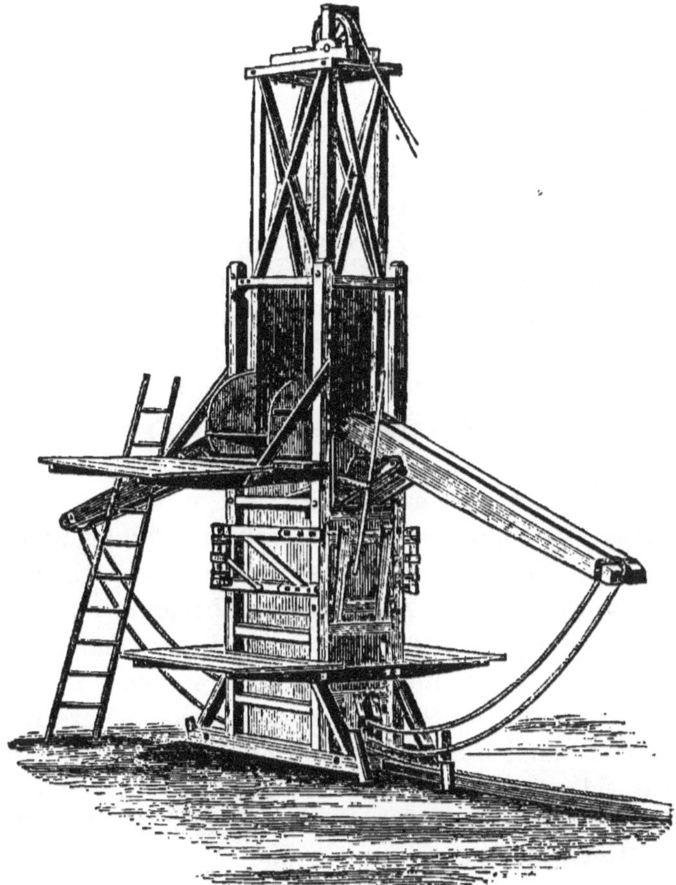

Dederick's Hay Press.

the pressure of every bale, Dederick's patent capstan (fig. 210) is employed with this press. The horse or horses, in passing around, wind up the rope on a horizontal wheel or drum. The possibility of any accident by slipping backwards is prevented by the pawl or anchor, *E*, at the end of the lever. When the

Fig. 210.

pressure is completed, the driver touches the upright rod, and detaches the wheel or drum, by which the rope is drawn backwards, without stopping the horses, which continue to walk around the circle.

This capstan answers an admirable purpose in using the common horse hay-fork, by obviating the necessity of backing up at every forkful.

The New York Beater Press Company, of Little Falls, manufacture a press, working like a pile engine, and reducing the hay to a degree of compactness nearly equal to that of solid wood. These bales are well adapted to long conveyance by land or shipment to foreign ports.

Hay Loaders.—Several of these, of different construction, have been tried to a limited extent, but, so far, the experiments have been but partially successful, or the machines have not proved themselves fully adapted to general use. Their expense, when compared with the horse-fork, and, to some degree, their cumbersome character, have proved objections. They mostly require very smooth meadows, are often difficult to work in the wind, and those constructed on the endless-rake principle are found to carry up small stones or gravel into barley, endangering the thrashing machine. Further ingenuity and labor on the part of inventors appear to be required, to place them generally within the reach of farmers.

CHAPTER XIII.

THRASHING, GRINDING, AND PREPARING PRODUCTS.

THRASHING MACHINES.

The old mode of beating out grain with the hand flail, (fig. 211,) has now nearly passed away, and thrashing machines have come into general use.

S. E. Todd makes the following statement relative to the saving of labor effected by these machines: "I have thrashed a great deal of grain of all kinds, with my own flail; and I have talked with others who have been accustomed to thrash their grain with flails, and I have come to the conclusion that the following figures represent a fair average as to the quantity of grain that an ordinary laborer will be able to thrash and clean in a day, viz.: Seven bushels of wheat, eighteen bushels of oats, fifteen bushels of barley, eight bushels of rye, and twenty bushels of buckwheat. In order to make this more intelligible, it will be necessary to double the number of bushels that one man is able to thrash, as two men will be required to clean the grain with a fanning mill.

Fig. 211.

An Old Flail.

"In order to labor economically and advantageously with a thrashing machine, two horses, at least, and three men are necessary. In most instances four or five men will be required, which will make a force equal to fifteen men with flails. Such a gang of hands, and two good horses, with such a thrasher and cleaner as Harder's, are capable of thrashing and cleaning of the same kind of grain to which allusion has been made, one hundred and seventy bushels of wheat, three hundred and twenty-five of oats, two hundred and twenty of barley, one hundred and eighty of rye, and two hundred and sixty of buckwheat. Some manufacturers of thrashing machines fix the average day's work higher than these figures. In some instances, I will acknowledge that a span of horses and five men can do much more than the amount represented by the foregoing figures; yet I am satisfied that in the majority of instances they will not thrash and clean a greater number of bushels than I have indicated. But, even at the low figures that I have recorded, such a machine as Harder's, or Palmer's Climax, or Wheeler, Melick & Co.'s, will be found to be a great labor-saving machine for thrashing all kinds of grain.

"There is one consideration that should not be overlooked in this estimate, which is the much greater amount of labor performed, with far less fatigue. When one laborer can perform the work of two or more workmen with less fatigue than has usually been required, a great point is gained."

J. Stanton Gould, estimating from a large number of statements that a saving of five per cent of the grain is effected by using the machine, over thrashing by the flail, computes the aggregate annual saving in the United States to be over eight million bushels of wheat, two million of rye, eight million of oats, and nearly a million of barley.

For farms of moderate size, the *endless-chain* powers for driving thrashing machines are most convenient, being

Fig. 212.

Endless-chain horse-power, driving a thrashing-machine.

compact or requiring but little room, easily conveyed from one place to another, and readily applicable to sawing wood, cutting straw, and to various other purposes. Fig. 212 represents a single horse-power, driving a small thrashing machine, with a simple, horizontal separator and straw carrier; and fig. 213 shows a two-horse power, (Emery's,) with the wheel-work on which the endless platform runs.

The power of these machines, and the amount of friction in running them, may be easily ascertained by the rule, already given in a former part of this work, for determining the power of the inclined plane; for the only

Fig. 213.

Two-horse Tread Power.

difference between the endless chain and a common inclined plane is, that in one the plane is fixed, and the body moves up its surface, and in the other the plane itself moves downward, and the weight or animal upon it remains stationary. The same principle applies in both cases.

First, to ascertain the friction, let the platform be placed on a level, with the horse upon it; then gradually raise the end until the weight of the horse will just give it motion. This will show the precise amount of the friction; for if the end be elevated one-twentieth of its length, then the friction is one-twentieth the weight of the horse and platform.

Secondly, to determine the power, when the end is still further raised, measure the difference between the height thus given and the length of the platform. If, for instance, the height of the inclination is one-eighth of its length, and the horse is found to weigh eight hundred pounds, then the power is one hundred pounds, or one-eighth the weight of the horse.

This rule will not, however, apply, when the *draught* of the horse is added to its weight; for it usually happens that the weight alone is not sufficient, without placing the platform in too steep a position for the horse to work comfortably. He is, therefore, attached to a whiffle-tree, and to ascertain the power requires the use of the dynamometer, in connection with the preceding mode.

Great improvement has been made of late years in the appendages of thrashing machines. The large number of laborers formerly employed in raking and separating the straw, and placing it on the stack, is now dispensed with, and the whole done by machinery, working by the same power that drives the thrasher.

Fig. 214.

Pitts' Thrasher and Straw Carrier.

Among the best and most widely known machinery for this purpose is that invented by H. A. Pitts, and represented in a portable form by fig. 214. It separates the grain, cleans it, and carries the straw by means of the elevator, (shown folded in the cut,) to the top of the highest stack.

The tread-power is successfully applied to churning, as shown in the cut, (fig. 215.) The employment of a sheep, of one of the larger breeds, has been found better and more convenient than a dog, as it is heavier, more quiet, less averse to the labor, and when the task is done, it is turned into the yard or pasture, where it is readily found next time.

The cost of horse-powers and thrashers combined, of the different forms, varies from $225 to $400.

CORN SHELLERS.

Fig. 215.

Churn worked by dog-power.

CORN SHELLERS are made for both hand and horse-power. One of the most convenient and compact hand machines is *Burrall's*, made of iron, furnished with a fly-wheel to equalize velocity, and worked by one person while another feeds it, one ear at a time. Or, one person alone will turn it with one hand, while the ears are dropped in with the other. Several other good corn shellers are made mostly of wood.

Fig. 217 represents a sheller, mostly of cast iron, driven by horses, by means of the band partly seen in the cut. The corn in the ear is thrown into the hopper at one end, and is separated from

Burrall's Corn Sheller.

the cob by rows of teeth revolving in a concave bed and

Fig. 217.

Horse-power Corn Sheller.

set spirally, thus carrying the cobs along and ejecting them from the opposite end.

For shelling corn in large quantities, powerful machines driven by horses or steam are required. An excellent sheller for this purpose is made by *Richards*, of Chicago. The corn is shoveled directly from the wagon or crib into the hopper, and requires no extra feeders, or hands, to keep the machine from choking. It is built wholly

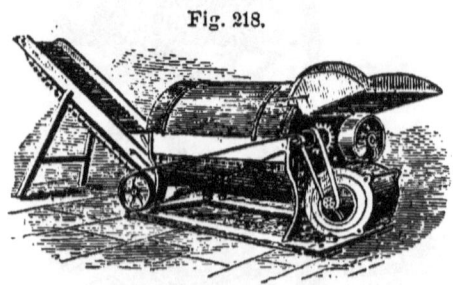

Fig. 218.

Richards' Corn Sheller.

of iron, combining strength and durability. The shellers are made of different sizes requiring from two to twelve horse-power. The former will shell one bushel, and the latter ten bushels per minute. The cost varies from $175 to $475. The following is a description of the working part.

The shelling cylinder is made of heavy rods of wrought iron, placed equidistant, presenting a corrugated surface, which cannot wear smooth. Within this, a revolving iron cylinder, with chilled teeth, thrashes the corn against the surfaces of the rod cylinder. The teeth approach the rods sufficiently close to keep every ear in rapid motion, shelling one ear or one bushel with the same facility. A regulator at the discharge end places the machine within control of the operator. The spaces between the rods allow the shelled corn to escape freely, thus lessening the draught, relieving the cylinder from clogging and from all liability to cut or grind the grain.

The cleaner consists of a cylindrical screen revolving around the whole length of the sheller, and extending beyond it. A heavy fan blast passes directly through this screen, and under it, subjecting the corn to two separate cleanings, and delivering it in good condition for market. The cleaned corn discharges upon either side desired, and the cobs are delivered with the dust at the end of the machine.

ARCHIMEDEAN ROOT-WASHER.

The spiral principle has been successfully applied in the *Archimedean Root-washer*, (fig. 219.) The roots to be

Fig. 219.

Croskill's Archimedean Root-washer.

washed are first delivered into a hopper, from which they

pass into an inclined cylinder made of strips of wood with grate-like openings. The cylinder has two portions separated by a partition, in the first of which they remain while the handle is turned for washing them. As soon as the washing is finished, the motion of the handle is reversed, which throws them into the other part, which has a spiral partition, along which they pass until they drop into a spout outside.

ROOT SLICERS.

These are mostly worked by hand. Out of a large number of inventions for this purpose, two are here represented;—Wellington's, made of iron, the knives being

Fig. 220. Fig. 221.

Wellington's Root Cutter.   *Root Cutter of Wood.*

on a revolving cone within the hopper, and shown below the machine; the other of wood, with the knives on the face of an iron wheel, which forms one side of the hopper. These two machines will cut roots at the rate of about a bushel per minute.

FARM MILLS.

These are made of iron, and with burr-stones. The former are cheaper, and answer a good purpose for grinding feed for domestic animals. The latter may be also used for grinding flour.

Figure 222 represents an iron farm mill manufactured by R. H. Allen & Co., of New York. The grinding surfaces are of chilled iron, so arranged as to be self-sharpening, and to last a long time without repairs. When necessary, new plates are readily inserted. The mill is driven by horse or other power, the band from which is seen in the cut. It will grind from five to ten bushels per hour, varying with the fineness of the meal and the amount of driving power. A two-horse railway power may be used to advantage.

Fig. 222.

Allen's Horse Mill.

This mill is about three feet square, four feet high, and weighs three hundred pounds. Several other iron mills of a similar character are made by different manufacturers, and usually cost about $50.

Among the burr-stone farm .mills, one of the best, most compact, and most substantial, is *Forsman's*, of Chicago,

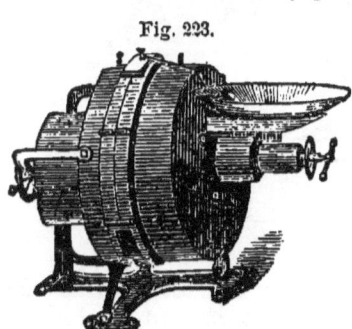

Fig. 223.

Forsman's Mill.

represented by fig. 223. It will be perceived from this

figure that the spindle is horizontal, and the face of the stones vertical. The frame is of iron. The diameters of the stones vary from sixteen to thirty inches, and the weight from 400 to 1,500 lbs. The smaller size may be run with a power of one to four horses; the larger, with that of ten to thirty horses. Prices $150 to $375. The manufacturers claim that it will grind from one and a half to three bushels per hour, for each horse-power used in driving it.

THE COTTON GIN.

Since the invention of the Cotton Gin by Eli Whitney, great improvements have been made, by which the cotton is cleaned with great rapidity and in a perfect manner.

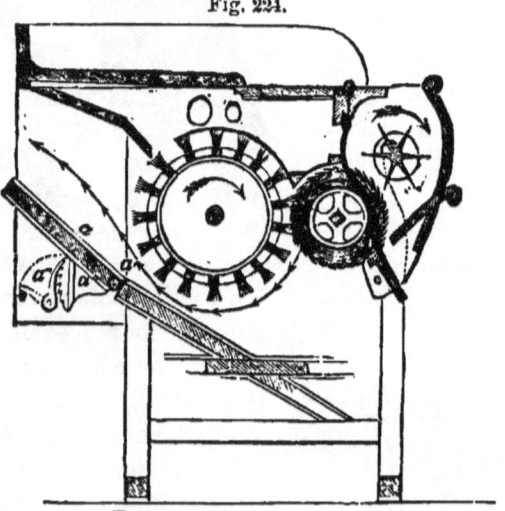

Fig. 224.

Emery's Cotton Gin—Section.

The machine manufactured by H. L. Emery, of Albany, is one of the best for this purpose. It is represented in section in fig. 224. The hopper, at the right, is furnished with what is termed a *Picker Roll Supporter*, which revolves within the hopper, in the direction shown by the arrow, and prevents the cotton from becoming packed. It is then taken by the teeth of the saw cylinder, which reduce the cotton to a fine condition. These teeth are swept by the brush cylinder, which, running in the same direction with the teeth, and slightly faster, carries the cotton off from them. Fig. 225 represents the operation, the seed escaping from the bottom of the hopper and the

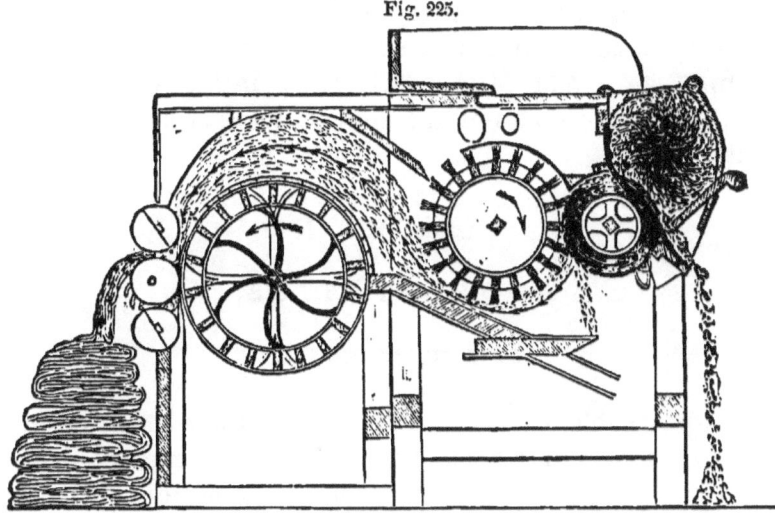

Fig. 225.

Emery's Cotton Gin, with Condenser.

cotton thrown to the rear into the condenser, which finishes the cleaning process and packs or condenses the lint cotton within a limited space. The arrows shown in the section indicate the direction of the revolutions of the picker, saws, and brush cylinder, and also the course which the cotton takes in passing through the gin and condenser.

PART II.

MACHINERY IN CONNECTION WITH WATER.

GENERAL PRINCIPLES.

HYDROSTATICS * treats of the weight and pressure of liquids when not in motion; HYDRAULICS,† of liquids in motion, as, conducting water through pipes, raising it by pumps, etc.; and HYDRODYNAMICS‡ includes both, by treating of the *forces* of the liquids, whether at rest or in motion.

CHAPTER I.

HYDROSTATICS.

UPWARD PRESSURE.

A remarkable property of liquids is their pressure *in all directions*. If we place a solid body, as a stone, in a vessel, its weight will only press upon the bottom; but if we pour in water, the water will not only press upon the bottom, but against the sides. For, bore a hole in the side, and the side pressure will drive out the water in a stream; or bore small holes in the sides and bottom of a tight wooden box, stopping them with plugs; then press this box, empty, bottom downward, into water, allowing none to run in at the top. Now draw one of the side plugs, and the water will be immediately driven into the

---

* From two Greek words, *hudor*, water, and *statos*, standing, or at rest.
† From two Greek words, *hudor*, water, and *aulos*, a pipe.
‡ From two Greek words, *hudor*, water, *dunamis*, power.

box by the pressure outside. If a bottom plug be drawn, the water will immediately spout up into the box, showing the pressure *upward* against the bottom. Hence the pressure in *all directions*, upward, sideways, and downward, is proved.

The upward pressure of liquids may be shown by pouring into one end of a tube, bent in the shape of the letter U, enough water to partly fill it; the upward pressure will drive the water up the other side until the two sides are level.

On this principle depends the art of conveying water in pipes under ground, across valleys. The water will rise as high on the opposite side the valley as the spring which supplies it. The ancient Romans, who were unacquainted with the manufacture of strong cast-iron pipes, conveyed water on lofty aqueducts of costly masonry, built level across the valleys. Even at the present day, it has been deemed safest to build level aqueducts for conveying great bodies of water, as in very large pipes the pressure would be enormous, and might result in violent explosions.

If the valleys are deep, the pipes must be correspondingly strong, because, the higher the head of water, the greater is the pressure. For the same reason, dams and large cisterns should be strongest at bottom. Reservoirs made in the form of large tubs require the lower hoops to be many times stronger or more numerous than the upper.

MEASUREMENT OF PRESSURE AT DIFFERENT HEIGHTS.

The amount of pressure which any given height of water exerts upon a surface below may be understood by the following simple calculation:

If there be a tube one inch square (with a closed end), half a pound of water poured into it will fill it to a height

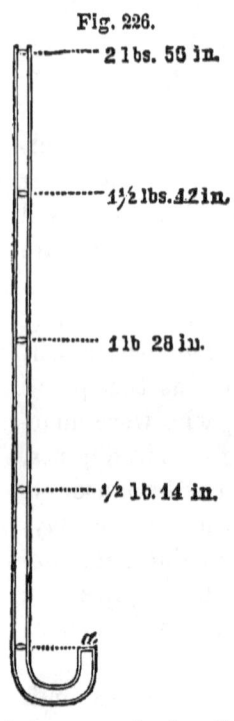

Fig. 226.

of fourteen inches;* one pound will fill it twenty-eight inches; two pounds, fifty-six inches; ten pounds, twenty-three feet; twenty pounds, forty-six feet, and so on. Now, as the side pressure is the same as the pressure downward for the same head of water, the same column will, of course, exert an equal pressure on a square inch of the *side* of the tube. Or, if the tube be bent, as shown in the annexed figure (fig. 226), the pressure upward on the end of the tube, at *a*, will be the same for the various heights.

Now, as the pressure of a column fifty feet high is about twenty-two pounds on a square inch, the pressure on the *four* sides is equal to eighty-eight pounds for one inch in length. Hence the reason that considerable strength is required in tubes which have much head of water, to prevent their being torn open by its force.

DETERMINING THE STRENGTH OF PIPES.

The question may now arise, and it is a very important one, How thick must be a lead tube of this size to prevent danger of bursting with a head of fifty feet, or of any other height? To answer it, let us turn to the table of the *Strength of Materials* in a former part of this work, where we find that a bar of cast lead one-fourth of an inch square will bear a weight of fifty-five pounds. If the

---

* This is nearly correct, for a cubic foot (or 1,728 cubic inches) of water weighs 62 lbs. Consequently, one pound will be 27.9 cubic inches, and will fill the tube nearly 28 inches high.

tube be only one-sixteenth of an inch thick, one inch of one of its sides will possess an equal strength, that is, will bear fifty-five pounds only, and the tube would consequently burst with fifty feet head. If one-tenth of an inch thick, the tube would just bear the pressure, and, to be safe, should be about twice as thick, or one-fifth of an inch. Half this thickness would be sufficient for twenty-five feet of water, and would require to be doubled for one hundred feet. A round tube, one inch in diameter, having less surface to its sides, would be about one-third stronger. A tube twice the diameter would need twice the thickness; or if less in diameter, a proportionate decrease in thickness might take place. If, instead of cast lead, milled lead were used, the tube would be nearly four times as strong, according to the table of the strength of materials already referred to.

SPRINGS AND ARTESIAN WELLS

result from the upward pressure of water. Rocks are usually arranged in inclined layers (fig. 227), and when

Fig. 227.

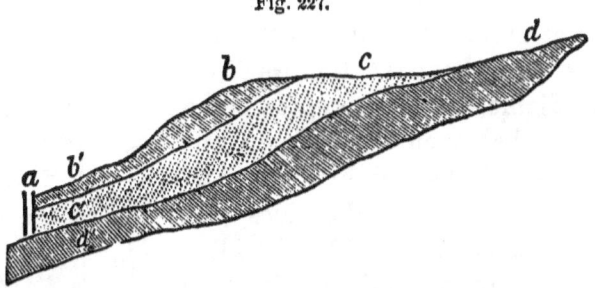

rain falls upon the surface, as at *c d*, it sinks down in the more porous parts between these layers, to *c*. If the layers happen to be broken in any place below, the water finds its way up through the crevices by the pressure of the head above, and forms springs. If there are no openings through the rocks, deep borings are sometimes made artificially, through which the water is driven up to the surface, as at *a*, forming what are termed *Artesian Wells*. The head of water which supplies them may be many miles distant, the place of discharge being on a lower level. It has sometimes been found necessary to bore more than a thousand feet downward before obtaining water which will flow out freely at the surface of the earth.

DETERMINING THE PRESSURE ON GIVEN SURFACES.

The pressure of liquids upon any given surface is always exactly in proportion to the height, no matter what the shape of the vessel may be. If, for instance, the vessel *a* (fig. 228), be one inch in diameter, and the vessel *b* be three inches in diameter, the water being equally high in both, the pressure on the whole bottom of *b* will be nine times as great as on the bottom of *a*; or any one inch of the bottom of *b* will receive as great a pressure as the bottom of *a*. Again, if the vessel *c*, broad at the top, be narrowed to only an inch in diameter at bottom, the pressure upon that inch will still be the same, most of the weight of its contents resting against the sides, *d d*.

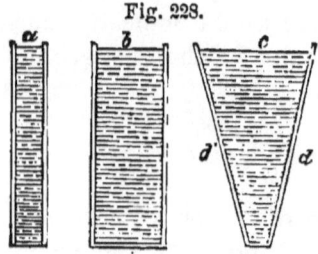

Fig. 228.

If the vessel, A (fig. 229), be filled with water to a height of fourteen inches, the pressure will be half a pound on every square inch of the bottom, or upon every square inch of the sides fourteen inches below the surface. If the tube, C, be an inch square, the water will be driven into it with a force of half a pound, and will press with that force against the one-inch surface of the stop-cock, C. If

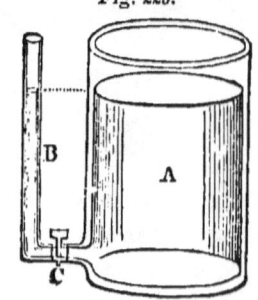

Fig. 229.

the tube, B, be now filled to an equal height, the same force will be exerted against the other side. To prove this, let the stop-cock be opened, when the two columns of water will remain at an exact level.

If enough water be now poured into the tube, B, to fill it to the top, it will immediately settle down on a level with the water in A, raising the whole surface in the latter. This result has seemed strange to many, who can not conceive how a small column of water can be made to balance a large one, and it has been therefore termed the *Hydrostatic Paradox*. But the difficulty entirely vanishes, and ceases to appear a paradox, when we remember that the water in the larger vessel rises as much more slowly than it descends in the smaller, as the large one exceeds the smaller; thus acting on the principle of virtual velocities in precisely the same manner that a heavy weight on the short end of a lever is upheld by a small weight on the long end. The great mass of water is supported directly by the bottom of A, in the same way that nearly all the weight on the lever is supported by the fulcrum.

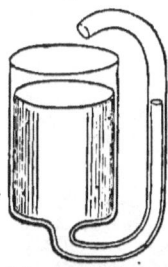
Fig. 230.
Attempted Perpetual Motion.

A man who was seeking a solution to the absurd mechanical problem of perpetual motion, and who supposed that the large mass in A would overbalance the small column in B, and drive it upward, constructed a vessel in the form shown in fig. 230, so that the small column, when forced upward, would flow back into the larger vessel perpetually. He was, however, greatly surprised to see the fluid in both divisions settle at the same level.

This principle may be further explained by the following experiment: A B (fig. 231) represents the inside of a metallic vessel, with a bottom, C, which slides up and down, water-tight. If water be poured in to fill the lower or larger part only, it will be found to press on the sliding bottom with a force exactly equal to its own weight; that is, if there is a pound of water, it will press on the bottom with a force equal to one pound. Now, if the bottom be pushed upward, so as to drive the water into the narrow part of the vessel, the pressure upon the bottom becomes instantly much greater, or equal to many pounds, the water being the same in quantity, but with a much higher head than before. Suppose the narrow part of the vessel is twenty times smaller than the larger part, then, in pushing the bottom up one inch, the water is driven twenty inches upward in the tube. So then, according to the rule of virtual velocities, it will require twenty times the force, because it moves upward twenty times faster.* This, then, is precisely similar to the instance where a pound on the longer end of a steelyard balances twenty pounds on the shorter end. In this instance, the upper parts, D D, of the vessel operate as the fulcrum of a lever, and offer resistance to the sliding part as soon as the water begins to ascend the tube.

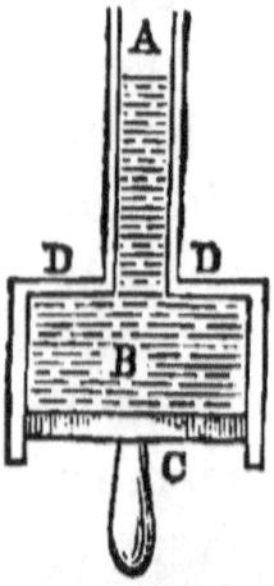

Fig. 231.

Fig. 232.

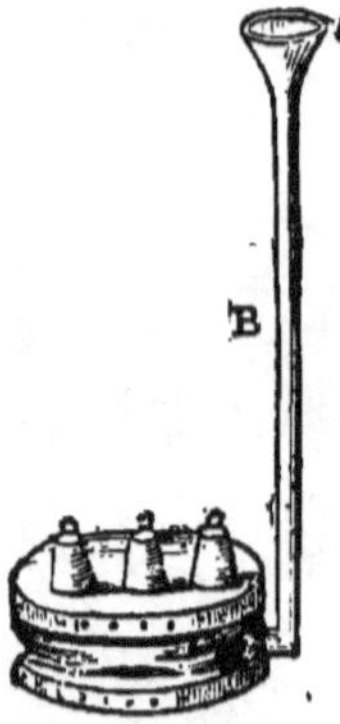

HYDROSTATIC BELLOWS.

This principle is shown in the *Hydrostatic Bellows* (fig. 232), which consists of two round pieces of board, connected by a narrow strip of strong leather; into it is inserted a long, narrow tube, B, with a small funnel, e, at the top. When water is poured into this tube, it will raise a weight as much greater than the weight of the

Hydrostatic Bellows.

---

* The pressure will be as great upon the bottom as if the vessel continued a uniform size all the way up.

water in the tube as the surface of the upper board exceeds the cross-section of the tube. Thus, if a pound of water fills a tube half an inch in diameter, and the bellows are two feet in diameter, then this pound will raise more than two thousand pounds on the bellows (if it be strong enough), because the surface of the bellows is more than two thousand times greater.

In the same way, a strong, iron-bound hogshead may be burst with the weight of a single gallon of water by pouring it into a long and narrow tube set upright in the bung of the filled hogshead. If, for instance, the inner surface of the hogshead be 20 square feet, or 2,880 square inches, a tube of water 23 feet high will press with a force of 10 pounds on every square inch, or equal to a force of 28,800 pounds, or 14 tons, on the whole surface.

HYDROSTATIC PRESS.

The *Hydrostatic Press* owes its extraordinary power to a similar principle; but, instead of a bellows, there is a moving piston in a strong metallic cylinder; and instead of being worked by the mere weight of the water, it is driven into the cylinder by means of the lever of a powerful forcing-pump. An instrument of this sort, possessing enormous power, was used to elevate the great tubular iron bridge in England. It was found necessary to make the sides of the cylinder into which the water was driven no less than eleven inches thick, of solid iron; and so great was the pressure given to the confined water, as to have forced it up through a tube higher than the summit of Mont Blanc. In the port of New York, vessels of a thousand tons' burden have been lifted by the hydrostatic press.

This machine has been applied in compressing hay, cotton, and other bulky substances into a compact form, so that they may occupy but little space, for conveyance to distant markets. The following figure (fig. 233) exhibits the different parts of this powerful machine. A is a cistern to supply water, which is raised by working the handle, B, of the forcing-pump; the water passes through the valve, C, opening upward, and through the spring valve,

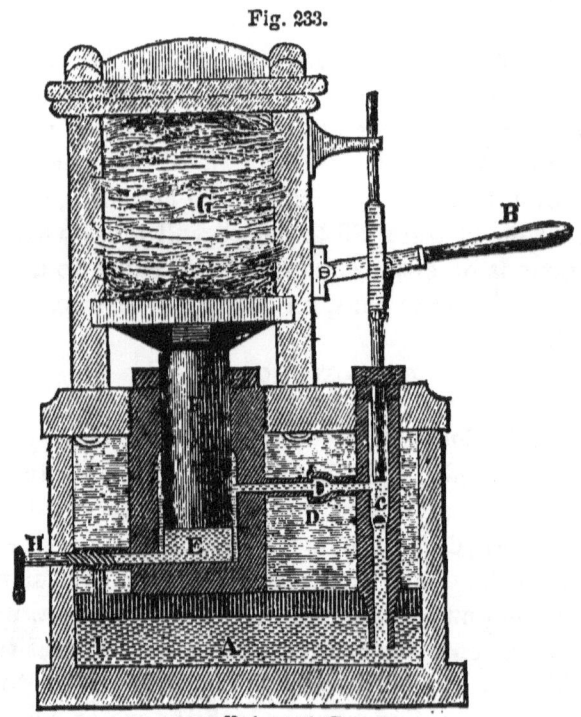

Fig. 233.

Hydrostatic Press.

D, opening toward the large cylinder, E. Being thus driven into the space, E, it raises the piston, F, and exerts a prodigious pressure upon the mass of hay or cotton, G. The piston is lowered by turning the screw, H, which allows the water to pass back into the cistern at I. In the figure the hay or cotton is shown as visible to the sight, in order to represent the whole more plainly; but in practice it is thrown into a square box or chamber of strong plank, of the size of the intended bundle. One side is hung upon stout hinges, and is opened for the removal of the bale when the pressing is completed.

To estimate the power of this machine, divide the square of the diameter of the piston, F, by the square of the diameter of the piston of the forcing-pump, and multiply the quotient by the power of the lever, B. For example, suppose the piston, F, is 16 inches in diameter, and the piston of the forcing-pump is 2 inches in diameter. The square of 16 is 256; divide this by 4, the square of 2, and the result will be 64. If the lever, B, increases the power five times, the whole power of the machine will be 320; that is, a force of one pound applied to the lever will raise the large piston with a force equal to 320 pounds; or, if a force of 100 pounds be given to the lever, the power will be 32,000 pounds, or 16 tons. Reducing the diameter of the smaller piston to half an inch, and increasing the force of the lever to twenty times, the whole power exerted will be thirty-two times as great, or equal to 960 tons. In ordinary practice, it is more convenient and economical to reduce the diameter of the larger piston to a few inches only, making the forcing-pump correspondingly small, the power depending entirely on the disproportion between them. Such presses may be worked rapidly by horse, water, or steam power.

One great advantage which the hydrostatic press possesses over those worked by screws results from the little friction among liquids, nearly the only friction existing in the whole machine being that of the two pistons, which is comparatively small. Another is the smallness of the compass within which the whole is comprised; for a man might, with one not larger than a tea-pot, standing before him on a table, cut through a thick bar of iron with as much ease as he could chip pasteboard with a pair of shears.

SPECIFIC GRAVITIES.

In connection with Hydrostatics, the subject of the specific gravities of bodies is one of importance. The specific gravity of a substance is its *comparative weight* with some other substance, an equal bulk of each being taken. Water is usually the standard for comparison.

To ascertain the specific gravity, weigh the body both in and out of water, and observe the difference; then divide the whole weight by this difference, and the quotient will be the specific gravity sought. For example, if a stone weighs 12 lbs. out of water and 7 lbs. in water, divide 12 by 5, and the quotient is 2.4, which shows that the stone is $2^4|_{10}$ times heavier than water. Figure 234 shows the mode of weighing the body in water, by suspending it beneath a balance on a hair or thread.

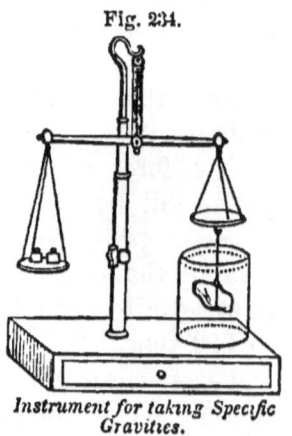

Fig. 234.

Instrument for taking Specific Gravities.

It was in a similar way that Archimedes is said to have succeeded in detecting the suspected fraud in the manufacture of the golden crown of the ancient king of Syracuse. He first weighed it, and then found that it displaced more water when plunged in a vessel just filled, than a piece of pure gold, and also that it displaced less than silver, whence he inferred the mixture of these two metals.

When the specific gravity of a substance lighter than water is to be ascertained, it is loaded down by a weight, so as to sink in water, for which allowance is made in the calculation. A very simple way to determine this in different kinds of wood is to form them into rods or sticks of uniform size throughout, and then to observe what portion of them sink when placed endwise in water.

TABLE OF SPECIFIC GRAVITIES.

A knowledge of the specific gravities of various substances becomes useful in many ways, among which is ascertaining the weight of any structure, machine, or implement, by knowing that of the material used in its manufacture; determining the cost, by the pound, of such material; or knowing the bulk or size of any load for a team. The latter may often be of great use in ordinary practice, by enabling the teamster to calculate beforehand the amount of load to give his horses, whether in timber, plank, brick, lime, sand, or iron, without first subjecting them to overstraining exertions in consequence of error in random guessing.

Tables of specific gravities, for this purpose, and weights of a cubic foot of different substances, are here given.

TABLE OF SPECIFIC GRAVITIES.

Metals.

Gold, pure............19.36	Iron............7.78
" standard............17.16	" cast............7.20
Mercury............13.58	Steel............7.82
Lead............11.35	Brass, common............7.82
Silver............10.50	Tin............7.29
Copper............8.82	Zinc............6.86

Stones and Earths.

Brick............1.90	Gypsum............1.87 to 2.17
Chalk............2.25 to 2.66	Limestone............2.38 to 3.17
Clay............1.93	Lime, quick............ .80
Coal, anthracite, about............1.53	Marble ....2.56 to 2.69
Coal, bituminous............1.27	Peat............60 to 1.32
Charcoal............ .44	Salt, common............2.13
Earth, loose, about............1.50	Sand............1.80
Flint............2.58	Slate............2.67
Granite, about............2.65	

Woods—dry.

Green wood often loses one-third of its weight by seasoning, and sometimes more. The same kind varies in compactness with soil, growth, exposure, and age of the trees.

Apple	.68 to .79	Pine, yellow	.55 to .66
Ash, white	.72 to .84	Oak, English	.93 to 1.17
Beech	.72 to .85	" white	.85
Box	.91 to 1.32	" live	.94 to 1.12
Cherry	.71	Poplar, Lombardy	.40
Cork	.24	Pear	.66
Elm	.58 to .67	Plum	.78
Hickory	.84 to 1.00	Sassafras	.48
Maple	.65 to .75	Walnut	.67
Pine, white	.47 to .56	Willow	.58

Miscellaneous.

Beeswax	.96	Oil, whale	.92
Butter	.94	" turpentine	.87
Honey	1.45	Sea water	1.02
Lard	.94	Sugar	1.60
Milk	1.03	Tallow	.93
Oil, linseed	.94	Vinegar	1.01 to 1.08

Weights of a Cubic Foot of various Substances, from which the Bulk of a Load of one Ton may be easily calculated.

Cast Iron	450	pounds.
Water	62	"
White pine, seasoned, about	30	"
White oak, " "	52	"
Loose earth, about	95	"
Common soil, compact, about	124	"
Clay, about	135	"
Clay with stones, about	160	"
Brick, about	125	"

Bulk of a Ton of different Substances.

23 cubic feet of sand, 18 cubic feet of earth, or 17 cubic feet of clay, make a ton. 18 cubic feet of gravel or earth before digging make 27 cubic feet when dug; or the bulk is increased as three to two. Therefore, in filling a drain two feet deep above the tile or stones, the earth should be heaped up a foot above the surface, to settle even with it, when the earth is shoveled loosely in. A cubic foot of solid half-rotted manure weighs about 56 lbs., requiring about 36 cubic feet to the ton. If coarse or dry, more will be required. Hay varies much in specific gravity with the kind, and the degree of pressure in the bay or stack; but good timothy hay, under medium pressure, requires about 500 cubic feet to the ton; clover, variable,—about one-half more.

CHAPTER II.

HYDRAULICS.

VELOCITY OF FALLING WATER.

Liquids in motion are subject to the same laws as solids in motion. Falling water increases in velocity at the same rate that the motion of falling solids is accelerated, as already explained under the head of *Gravitation*. Thus a perpendicular stream of water descends one foot in a quarter of a second, four feet in half a second, nine feet in three-fourths of a second, and sixteen feet in one second. Like falling solids, the velocity at the end of the first quarter will be eight feet per second; at the end of the second quarter, sixteen feet; at the end of the third quarter, twenty-four feet; and at the end of the fourth quarter, thirty-two feet per second.

Now, if there be an *orifice* made in the side of a vessel of water, the water will spout out with the same swiftness as if it fell perpendicularly from an equal height, were it not retarded a little by friction. For example, if the head of water is one foot above the orifice, the velocity would be at the rate of eight feet per second, but for friction, which reduces it to about five and a half feet. The velocity for any other height of head may be easily found by deducting the same proportionate rate from the velocity of a falling body. Thus, for example, if the head be sixteen feet, the speed would be thirty-two feet (as shown under *Gravitation*), from which, deducting the friction, the real velocity would be about twenty-two feet per second.

It has been already shown that the velocity of a falling body increases at the same rate as the increase in the time of falling; for instance, the speed is twice as great in two seconds as in one; three times as great in three seconds;

four times as great in four seconds; and so on. But the *distance fallen through* increases as the *square* of the time; that is, it is four times as great in two seconds, nine times as great in three seconds, sixteen times as great in four seconds, etc. Thus we see that, in order to produce a twofold velocity, a fourfold height is necessary, etc. So also in the escape of water under a head: to double the velocity of the stream, the head must be four times as high; to triple it, the head must be nine times as high, etc.

DISCHARGE OF WATER THROUGH ORIFICES AND PIPES.

The discharge of water from a vessel is greatly influenced by the nature of the orifice through which it flows.

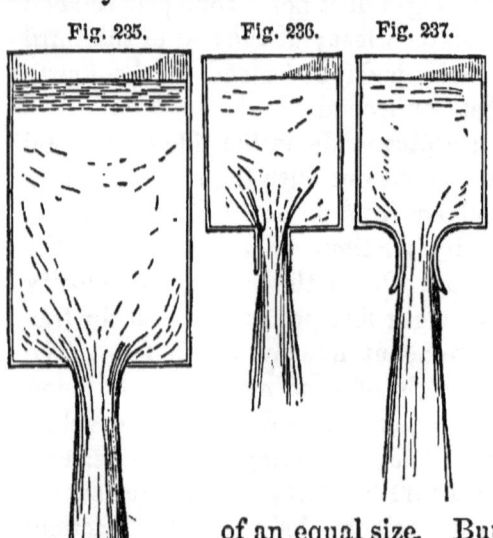

Fig. 235.   Fig. 236.   Fig. 237.

If, for example, a vessel or cistern have a thin bottom of tin, with a smooth, circular hole, we might naturally suppose that the discharge would be as easy as it could be made, and that water would pass as rapidly through it as through any orifice of an equal size. But this is not the fact. As the particles approach this orifice, their motion throws them across, and they partly obstruct the opening; it will be seen that they converge toward a point just under the orifice, where the stream will be considerably contracted (Fig. 235). If a short tube be inserted into the hole (the head being the same), this crossing of particles will be partly prevented, and the liquid will flow more rapidly. The greatest effect is produced when the tube is twice as long as its diameter (fig. 236). If the tube be enlarged at its upper and lower end, similar to the form of the contracted stream of water in fig. 237, the quantity discharged is greatly increased.

When water flows down an inclined plane, the same law applies as to the motion of a solid body rolling down a plane. The velocity increases as the square of the distance, and is the same as the velocity of a body falling freely downward from a height equal to the perpendicular height of the plane. Unless the stream, however, is very large, its speed is quickly diminished by the friction of its channel,* until this friction becomes as great as the descending force, after which the motion becomes uniform. Hence the reason that large streams, with an equal degree of descent, flow so much more rapidly than small ones, the gravitating force being so much greater that friction has a less retarding effect upon them.

In pipes which wholly surround the flowing stream, the friction becomes still greater, and the difficulty is only obviated by making the pipe of larger dimensions than would otherwise be necessary, so as to allow a free passage of a sufficient quantity of water through the centre of the tube, while a ring or hollow cylinder of water is nearly at rest all around it. The tables in the Appendix exhibit this decreased velocity in tubes of various sizes.

Lead pipes, for conveying the water of springs underground, should commonly be three-fourths of an inch in diameter. Five-eighths will answer where the distance is short and the descent considerable. But with a half-inch pipe, the friction of the sides is so great, compared with the small force of the current, that but little water will flow through it.

---

* Which increases as the square of the velocity.

VELOCITY OF WATER IN DITCHES.

It is often of great practical utility to know what amount of water may be carried off in draining, or supplied in irrigation, by channels of any given size and descent. The following rule will apply to all cases, from the plow-furrow to the mill-race, or even to the large river, and may be used by any boy who understands common arithmetic, and which is illustrated and made plain by the example that follows the rule.

To ascertain the mean (or average) velocity of water in a straight channel of equal size throughout:

Let f = the fall in two miles in inches;
Let d = the hydraulic mean depth;
Let v = the velocity in inches per second;
then the rule is thus expressed: $v = 0.91 \sqrt{fd}$, or, in plain words, the velocity is equal to the hydraulic mean depth, multiplied by the fall, with the square root of this product extracted, and then multiplied by 0.91.

The "*hydraulic mean depth*" is found by dividing the cross-section of the channel by the *perimeter*, or border. The perimeter is the aggregate breadths of the sides and bottom of the channel.

The rule will be rendered quite plain by an example. Suppose a smooth furrow is cut six inches wide and four inches deep, with perpendicular sides, and that it descends one inch in a rod; to find the quantity of water that will flow through it. One inch fall in a rod is 320 inches in a mile, or 640 in two miles. The perimeter in contact with the water will be six inches on the bottom and four inches in each side = 14 inches. The area of the cross-section will be six times 4 = 24, which, divided by 14, the perimeter, gives 1.7 = the hydraulic mean depth. Then, by applying the preceding rule:

$v = 0.91 \sqrt{640 \times 1.7}$, or $v = 0.91 \times 33 = 30$ inches, is the velocity per second, which would be about three gallons per second, or three hogsheads per minute.

An open ditch, therefore, with smooth sides, conveying a stream of this size, would carry off, in one hour, from an acre of land, all the water which might fall by half an inch of rain, during the wet season; for half an inch of rain would be one hundred and eighty hogsheads per acre, which would pass off in one hour; or it would supply in one hour, by the process of irrigation, as much water as a heavy shower of half an inch. Where the descent is greater, the increased quantity may be readily calculated by the rule given. The capacity of smooth-sided underground channels may be determined in the same way; but if built of rough stones, great allowance must be made, as they will retard the flow of water.

In common practice, too, even with straight, open ditches, the velocity will be much diminished by the rough sides.

LEVELING INSTRUMENTS.

The simplest mode of *leveling*, or ascertaining the slope for ditches, is to cut a few yards of the ditch, so that water may stand in it, and then to set two sticks perpendicularly, both rising to an equal height above the surface.

Fig. 238.

Simple Method of Taking Levels.

The sticks should be measured at equal distances from the top downward, and marked, and then pressed into the earth, till the water reaches the mark. The level may then be determined with much accuracy by "sighting" over the tops of these sticks. Fig. 238 exhibits this arrangement. The shorter the sticks, and the longer the piece of water, the less will be the liability to error.

A leveling instrument for general use, sufficiently accurate for all common purposes, may be made in the following manner:—Procure a mason's or carpenter's spirit-level, (*a*, fig. 239) fasten it to the upper side of a straight bar of wood, and on the ends of this bar secure small, upright pieces of tin plate, *b, b*, having openings cut in the centre. Horizontally across these openings, draw very fine wire, which shall be exactly at equal heights above each end of the bar of wood, to adjust which, one should be capable of sliding, and be screwed or wedged to its place at pleasure. The bar is placed on a common compass staff, as shown in the cut, turning at the ball and socket, below the spirit level. A tripod (fig. 240) is better, if it can be had. A small spirit-level should be secured across the bar, so that it may be adjusted both ways. When the bar is made perfectly level, as shown by the air-bubble, by sighting through at the two wires, the levelness or descent of the land may be determined.

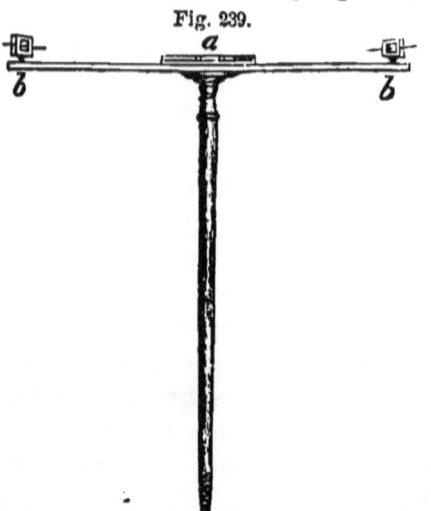

Fig. 239.

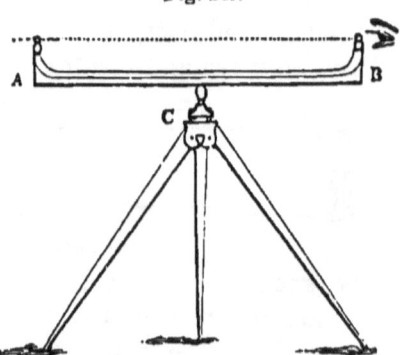

Fig. 240.

To ascertain whether these threads are both of equal heights above the bar, let a mark be made where they intersect some distant object; then reverse the instrument, or turn it end for end, and observe whether the threads cross the same mark. If they do, the instrument is correct; but if they do not, then one of the sights must be raised or lowered until it becomes so.

In laying out canals and rail-roads, where extreme accuracy is needed, the *spirit-level*, attached to a telescope, is used. So great is the perfection of this instrument, that separate lines of levels have been run with it for sixty miles, without varying two-thirds of an inch for the whole distance.

The use of a cheap and simple instrument to determine the position and descent of ditches with ease and precision, before commencing with the spade, will save a vast amount of the trouble and expense which those often meet with whose only method is to " *cut and try.*"

HYDRAULIC MACHINES.

ARCHIMEDEAN SCREW.

Machines for raising water are of frequent use on every farm. One of the simplest contrivances, in principle, for this purpose, is *the Screw of Archimedes.* It may be

Fig. 241.

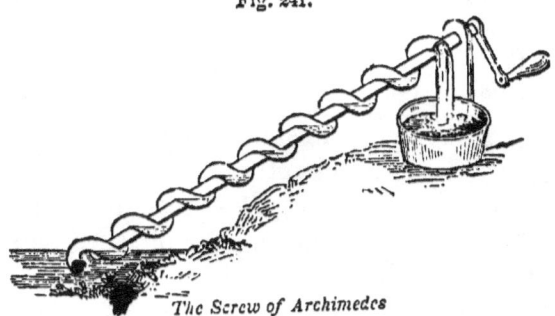

The Screw of Archimedes

easily made by winding a lead tube around a wooden cylinder or rod (fig. 241), in the form of a screw. When

placed in an inclined position, with one end in water, and made to revolve, the water resting at the lower side of each turn of the screw is gradually carried from one end to the other, and discharged at the upper extremity. Its simplicity and small liability to get out of order render the Archimedean screw sometimes useful where water is to be raised from an open stream to a short distance, as for irrigation, the motion being easily imparted to it by means of a small water-wheel, driven by the stream.

PUMPS.

Great improvement has been made in the *common pump* for farms within a few years. The best cast-iron pumps, made almost wholly of this metal, exceed in durability and ease of working those formerly constructed of wood, and excel others in cheapness. Fig. 242 exhibits the working of the common pump, the water first passing through the fixed valve below, and then through the one in the piston; both opening upward, it cannot flow back without instantly shutting them. The water is driven up by the pressure of the atmosphere, explained in the next chapter.

Fig. 243 is an iron cistern pump, showing the mode of bolting it to the floor or platform, and representing, also, its neat and compact form, occupying but little space at one side, or in the corner of a kitchen, over the cistern.

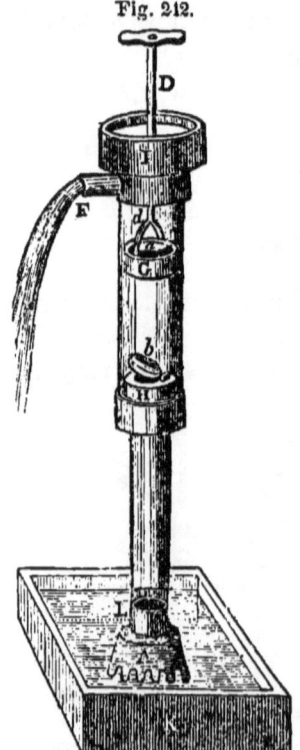

Fig. 242.

Common Pump: b, lower or fixed valve, G, piston with valve, a, opening upward; D d, piston-rod; F, spout.

Fig. 244 represents a cistern or well pump, so constructed that the working parts are about 20 inches below the platform, or base of the pump, and it is therefore well adapted for outdoor work. If the well or cistern is kept covered tight, the pump will not freeze below the platform. It will succeed in any well not over twenty feet deep, and by means of its various couplings may be made to draw water in a horizontal or inclined position, provided the whole height is not much over twenty feet.

Fig. 244.

Cistern Pump.   *Non-freezing Pump.*

An excellent deep-well pump, made by Cowing & Co., of Seneca Falls, N. Y., is represented by fig. 245; the working part, being placed at the bottom of the well, is adapted to any depth of water, the rod working safely within the pipe. The lower part of the cylinder is furnished with a strainer, and is plugged at the bottom, to

220   MACHINERY IN CONNECTION WITH WATER.

Fig. 246.

Fig. 245.

Drive Pump.

Deep-well Pump.

prevent the ingress of sand and mud. The connecting pipe between the cylinder at the bottom and the standard at the top is wrought or galvanized iron. The pump, of course, needs bracing, to prevent swinging when worked.

Drive Pumps.—Fig. 246 represents the new mode of making wells, by simply driving into the earth common iron gas-pipe, pointed at the lower end, and perforated at the sides, near the lower extremity, for the ingress of water—thus obviating entirely the cost and labor of digging wells. If driven through a subterranean spring, a stratum of water, or a wet layer of sand or gravel, it is obvious that the water will immediately flow through the perforations into the pipe; and, by attaching a good pump to the pipe and pumping for a time, all the particles of sand and fine gravel will be drawn out; and the cavity thus formed around the perforations will remain filled with pure water. These tubes and pumps are admirably adapted to localities where large beds of wet gravel exist fifteen or twenty-five feet below; and, in fact, to all soils where large stones are not abundant. Where these occur, the pipe must be withdrawn, and tried in a new place, until success is attained.

In the *Chain Pump*, a partial cross-section of which is

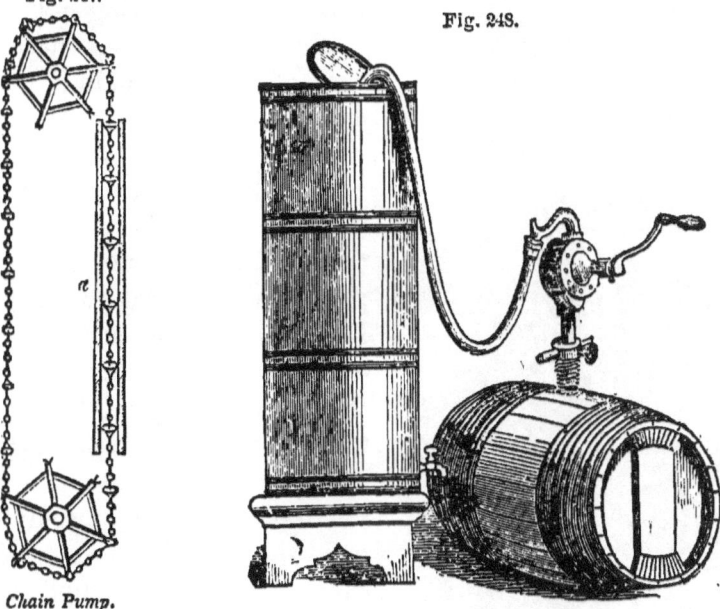

Fig. 247.

Fig. 248.

Chain Pump. Section.

Rotary Pump, for Barrels, etc.

here shown, (fig. 247), the chain is made to revolve rapidly on the angular wheel by means of a winch attached to the upper one, and being furnished with a regular succession of metallic discs, which nearly fit the bore in the tube, *a*, the water is carried up in large quantities. When the motion is discontinued, the water settles down again into the well, and consequently this pump is not liable to accident by freezing. By sweeping rapidly through the water, it preserves it in better condition, and prevents stagnation. The friction being very small, it will last a long time without wearing out.

Fig. 249.

Suction and Forcing-pump.

Rotary Pumps.—A succession of cavities made in the exterior of a short cylinder receive the water from the pump-tube below, and force it away into the elevating tube. When driven fast, it pumps with great rapidity. It possesses this advantage over the common pump, that the motion being continuous, no force is lost by repeatedly checking the momentum. In the figure on the preceding page, the pump is represented as inserted in a barrel of oil, which is to be emptied into the reservoir above, and is worked by hand. Larger rotary pumps are driven by horse and steam power.

TURBINE WATER-WHEELS. 223

Suction and Forcing-Pump.—The accompanying cut (fig. 249) represents a suction and forcing-pump combined in one, for the purpose of drawing water from a well or cistern, and forcing it to tanks in upper stories, or throwing water into upper rooms in case of fire. By lengthening the rod, the working parts may be placed at the bottom of a deep well, and the whole used as a deep well pump.

TURBINE WATER-WHEELS.

The large wooden wheels formerly used for the application of water power to mills and other machinery are rapidly giving place to iron *Turbine* wheels. Overshot wheels, the best kind formerly employed, were turned by the weight of the water, the whole of which was held in the slowly descending buckets of the wheel. Turbine wheels do not hold the water, but merely receive and impart the force of the rushing current, the water being held by the flume above. Hence, a turbine wheel of quite small size may impart to machinery nearly the whole force of a powerful current of water.

Turbine wheels are placed in a horizontal position, with vertical axes. Being under water, they never freeze; and they are not impeded by back-water when a flood occurs. There are two principal kinds among those in common use, — those, like the Reynolds wheel, which have a single opening at the side, through which the water is admitted; and such as the Leffel and Van de Water wheels, into which the water is admitted through several openings around them.

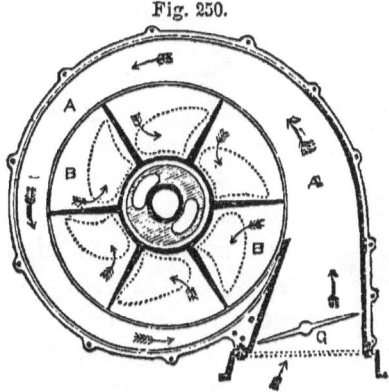

Section of Reynolds' Wheel.

Fig. 250 is a section of the Reynolds wheel; *G*, the gate for admitting the water through the horizontal shute from the flume; *A, A*, the circular passage for the water, which is gradually diminished in volume as it strikes the buckets or blades, *B, B*, and escapes through the bottom and top of the wheel. The arrows show the currents, and the curved dotted lines the openings through which the water escapes—the curved arrows exhibiting the rebounding of the current against the blades, before passing out through the issues. Fig. 251 is an exterior view of the wheel, showing the gate for the admission of the water; and fig. 252 represents the shaft and buckets separate.

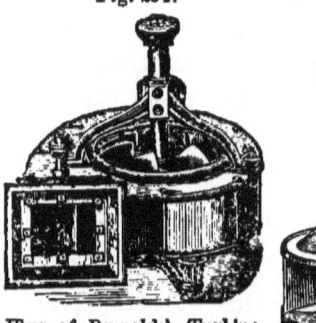

View of Reynolds' Turbine Wheel.

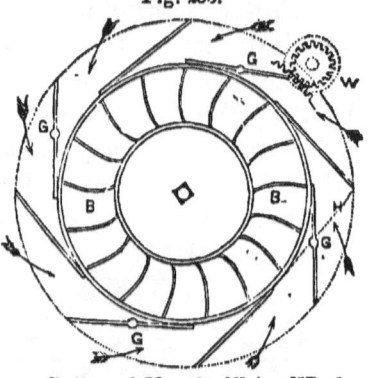
Section of Van de Water Wheel.

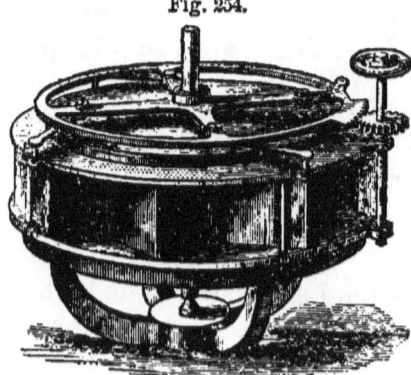
Van de Water Wheel.

Fig. 253 is a section of the Van de Water wheel, *G, G, G, G*, being the gates for admitting water, and *B, B*, the buckets — the arrows representing the entering currents. *H* shows, by dotted lines, the position of one of the gates when closed. The water, after entering the buckets, passes out below, where the blades are curved backwards, to receive all the force of the escaping water. Fig. 254 is a view of this wheel, showing the admission gates, and the wheel at the top, for opening and shutting the gates at one movement.

The Reynolds wheel is placed under water, outside the flume, and the current admitted at the side, as already stated. The Van de Water wheel is placed within, and on the bottom of the flume, in the floor of which a circular hole is cut, through which the water escapes. Both are excellent wheels, and are among those most extensively manufactured in the country—the former by George Tallcot, of New York, and the latter by the inventor, H. Van de' Water, of Attica, N. Y.

Turbine wheels, of the best construction, do not lose more than one-seventh or one-eighth of the whole descending force of the water. Hence, the power of any stream may be determined beforehand with much accuracy, if the descent or head and the number of cubic feet of water per minute are known. It has been already shown in this work, that a single horse-power is equal to lifting 33,000 lbs. one foot, per minute. This is equivalent to raising 530˙cubic feet of water to the same height, or 53 cubic feet, ten feet high. A stream, then, which falls 10 feet, and discharges 53 cubic feet in a minute, or nearly 1 per second, has an inherent force of one horse-power. Add one-seventh, making it about 60 cubic feet, and we have the size of a stream for one horse-power, at ten feet fall. Twenty feet descent would double the power, forty, quadruple it, and so on; and a similar increase result from employing a larger stream. As examples, a small wheel, seven or eight inches in diameter, will be sufficient for such a purpose. One of this size, with 20 feet head, and discharging 70 or 80 cubic feet of water per minute, will possess about three horse-power; and with forty feet head, requiring over 100 cubic feet per minute, it will have a power of eight or nine horses.

10*

The simple rule given in the second paragraph of the present chapter, for determining the velocity of a current of water spouting out under any given head, will enable any one who understands arithmetic to calculate the proper speed of a turbine wheel, which varies with the head and the diameter of the wheel. It is found that the buckets or blades should move with about two-thirds the velocity of the current as it rushes from the flume; hence, as an example, under a head of 16 feet, which drives out a stream about 22 feet per second, the exterior of the turbine wheel should move about 14 or 15 feet per second. If 1 foot in diameter, it should therefore revolve five times per second; or, if $2\frac{1}{2}$ feet in diameter, only twice per second. Other examples may be readily computed.

There are occasional opportunities for employing water power for driving farm machinery—as thrashing machines, mills for grinding feed, corn shellers, wood saws, straw cutters, etc., by bringing streams along hill-sides, or over bluffs; in which cases, turbine wheels would be cheaper than steam-engines, and require neither food nor fuel. The water of small streams might be saved in dams or ponds, giving a power of five or six horses for one day in each week for grinding, thrashing, and other purposes.

THE WATER-RAM.

One of the most ingenious and useful machines for elevating water is the *Water-ram*. It might be employed with great advantage on many farms, were its principle and mode of action more generally understood. By means of a small stream, with only a few feet fall, a current of water may be driven to an elevation of 50 feet or more above, and conveyed on a higher level to pasture-fields for irrigation, to cattle-yards for supplying drink to domestic animals, or to the kitchens of dwellings for culinary purposes.

THE WATER-RAM. 227

Its power depends on the momentum of the stream. Its principal parts are the reservoir, or air-chamber, *A*, (fig. 255), the supply pipe, *B*, and the discharge pipe, *C*. The running stream rushes down the drive, or supply-pipe, *B*, and, striking the waste valve, *D*, closes it. The stream being thus suddenly checked, its momentum opens the valve, *E*, upward, and drives the water into the reservoir, *A*, until the air within, being compressed into a smaller space by its elasticity, bears down upon the water, and again closes the valve, *E*. The water in the supply-pipe, *B*, has, by this time, expended its momentum, and stopped running; therefore the valve, *D*, drops open again, and permits it to escape. It recommences running, until its force again closes the waste valve, *D*, and a second portion of water is driven into the reservoir as before, and so it repeatedly continues. The great force of the compressed air in the reservoir drives the water up the discharge-pipe, *C*, to any required height or distance.

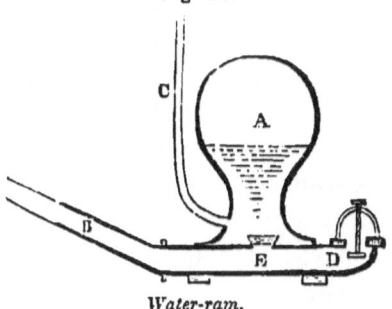

Fig. 255.

Water-ram.

The mere *weight* of the water will only cause it to rise as high as the fountain head; but like the momentum of a hammer, which drives a nail into a solid beam, which a hundred pounds would not do by pressure, the *striking force* of the stream exerts great power.

The discharge pipe, *C*, is usually half an inch in diameter, and the supply-pipe should not be less than an inch and a fourth. A fall of three or four feet in the stream, with not less than half a gallon of water per minute, with a supply-pipe forty feet long, will elevate water to a height as great as the strength of common half-inch lead pipe will bear.* The greater the height, in proportion to the fall of the stream, the less will be the quantity of water elevated, as compared with the quantity flowing in the stream, or escaping from the waste valve.

H. L. Emery gives the following rule for determining the quantity of water elevated from a stream:—Divide the elevation to be overcome by the fall in the drive-pipe, and the quotient will be the proportion of water, (passing through the drive-pipe), which will be raised,—deducting, also, for waste of power and friction, say one-fourth the amount. Thus, with 10 feet fall, and 100 feet elevation, one-tenth of the water would be raised if there were no friction or loss; but, deducting, say one-fourth for these, seven and a half gallons in each hundred gallons would be raised, the rest escaping, or being required to accomplish this result. Or, if the fall of the water in the supply-pipe be 3 feet, and the elevation required in the discharge-pipe be 15 feet, about one-seventh part of all the water will be elevated to this height of 15 feet. But if the desired height be 30 feet, then only about one-fourteenth part of the water will be raised; and so on in about the same ratio for different heights. A gallon per minute from the spring would elevate six barrels five times as high as the fall, in twenty-four hours, and at the same rate for larger streams. With a head of 8 or 10 feet, water may be driven up to a height of 100, or even 150 feet, provided the machine and pipes are strong enough. The best result is obtained when the length of the drive-pipe and the momentum it produces are just sufficient to overcome the reaction caused by the closing of

---

* When water is raised to a considerable elevation by means of the water-ram, the reservoir must possess great strength. If the height be 100 feet, the pressure, as shown on a former page, is about forty-four pounds to the square inch. With an internal surface, therefore, of only 2 square feet, the force exerted by the column of water, tending to burst the reservoir, would be equal to more than twelve thousand pounds.

the waste valve at each pulsation, and prevent the current of water from being thrown backward or up the drive-pipe; hence, the greater the disproportion between the fall and the required elevation, the longer or larger must be the drive-pipe, in order to obtain sufficient momentum. A descent of only a foot or two is sufficient to raise water to moderate elevations, but the drive-pipe should be of large bore. This pipe should always be very nearly straight, so that the water, by having a free course, may acquire sufficient momentum to compress the air in the ram, and push the water up the discharge-pipe. Water may be carried to a distance of a hundred rods or more, but as there is some friction in so long a discharge-pipe, a greater force is required than for short distances. The discharge-pipe should, therefore, be larger, as the length is increased. Half an inch diameter is a common size, but long pipes may be five-eighths or three-fourths; and, when practicable, it is more economical to reach an elevation with a short and strong pipe, and to use a lighter and weaker one for the upper part. A pit, lined with brick or smooth stone, for placing the ram, protects it from freezing; and both pipes should be under ground for the same reason. The supply or drive-pipe is usually 40 to 50 feet long; but where the fall is 8 or 10 feet, it should be sixty or seventy feet.

Unlike a pump, there is no friction or rubbing of parts in the water-ram, and, with clean water, it will act for years without repairs, continuing through day and night its constant and regular pulsations, unaltered and unobserved. A small quantity of sand, or of dead grass or other fibre, in the water, will be liable to obstruct the valves, and render frequent attention necessary.

WATER-ENGINES,

including those for extinguishing fires and for irrigating gardens, are constructed on a principle quite similar to

230 MACHINERY IN CONNECTION WITH WATER.

Garden-engine.

that of the water-ram. Instead, however, of compressing the air, as in the ram, by the successive strokes of a column of running water, it is accomplished by means of a forcing-pump, driving the water into the reservoir, from which it is again expelled with great power, by means of the elasticity of the compressed air. Fig. 256 represents a garden-engine, movable on wheels, which may be used for watering gardens, washing windows, or as a small fire-engine. Fig. 257 is another, of smaller size, for the same purposes, and in a neat and compact form, the working part being within the cylindrical case.

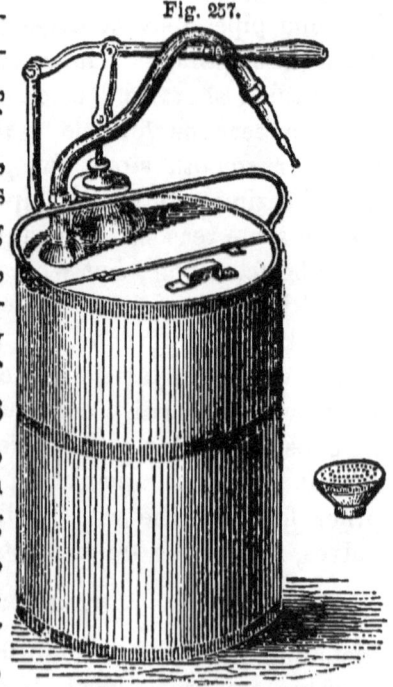

Cylindrical Garden-engine.

THE FLASH-WHEEL

is employed with great advantage where the quantity of water is large, and is to be raised to a small height, as in draining marshes and swamps. It is like an undershot wheel with its motion reversed; in fig. 258 the arrows show the direction of the current when driven upward. It must, of course, be made to fit the channel closely, without touching and causing friction. In its best form, its paddles incline backward, so as to be nearly up-

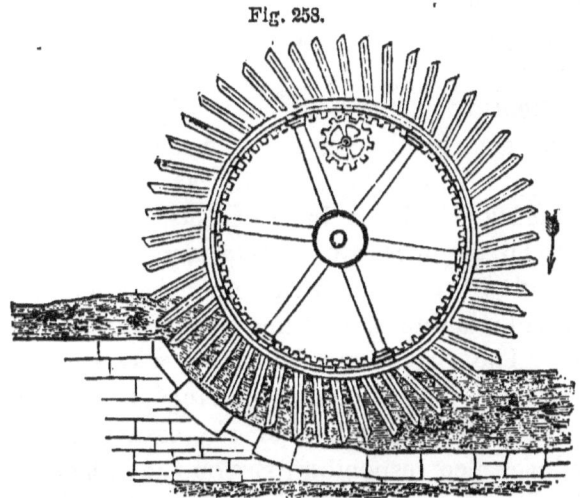

Fig. 258.

Flash or fen wheel for raising water rapidly short distances.

right at the time the water is discharged from them into the upper channel. It has been much used in Holland, where it is driven by wind-mills, for draining the surface-water off from embanked meadows. In England, it has been driven by steam-engines; and in one instance, an eighty-horse-power engine, with ten bushels of coal, raised 9,840 tons of water six feet and seven inches high, in an hour. This is equal to more than 29,000 lbs. raised one foot per minute by each horse-power, showing that very little force is lost by friction in the use of the flash-wheel.

WAVES.

NATURE OF WAVES.

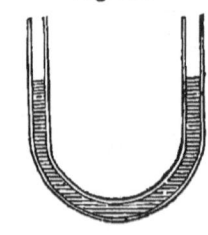

Fig. 259.

An inverted syphon, or bent tube, like that shown in fig. 259, may be used to exhibit the principle on which depends the motion of the waves of the sea. The action of the waves on shores and banks, and the inroads which they make upon farms situated on the borders of lakes and large rivers, present an interesting subject of inquiry.

If the bent tube (fig. 259) be nearly filled with water, and the surface be driven down in one arm by blowing suddenly into it, the liquid will rise in the other arm. The increased weight or head of this raised column will cause it to fall again, its momentum carrying it down below a level, and driving the water up the other arm. The surfaces will, therefore, continue to vibrate until the force is spent. The rising and falling of waves depend on a similar action. The wind, by blowing strongly on a portion of the water of the lake or sea, causes a depression, and produces a corresponding rise on the adjacent surface. The raised portion then falls by its weight, with the added force of the wind upon it, until the vibrations increase into large waves.

THE WATER NOT PROGRESSIVE.

The waves thus produced have a progressive motion (for reasons to be presently shown), as every one has observed. A curious optical deception attending this advancing motion has induced many to believe that the water itself is rolling onward; but this is not the fact. The boat which floats upon the waves is not carried forward with them; they pass underneath, now lifting it on their summits, and now dropping it into the hollows between. The same effect may be observed with the water-fowl, which sits upon the surface. It often happens, indeed, that the waves on a river roll in an opposite direction to the current itself.

If a cloth be laid over a number of parallel rollers, so far apart as to allow the cloth to fall between them, and a progressive motion be then given to them, the cloth remaining stationary, a good representation of waves will be afforded, and the cloth will appear to advance; or if a strip of cloth be laid on a floor, repeated jerks at one end will produce a similar illusion.

It is only the *form* of the wave, and not the water which composes it, which has the onward motion. Let the dark line in fig. 260 represent the surface of the water.

Fig. 260.

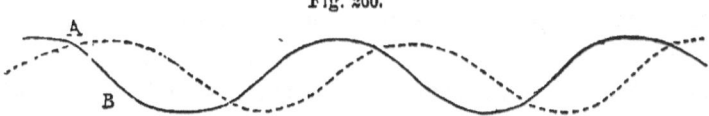

A is the crest of one of the waves, and being higher than the surface at *B*, it has a tendency to fall, and *B* to rise. But the momentum thus acquired carries these points so far that they interchange levels. The same change takes place with the other waves, and the dotted line shows the newly formed surface as the water thus sinks in one place and rises in another. The same process is again repeated, and each wave thus advances further on, and its progressive motion is continually kept up.

BREADTH AND VELOCITY OF WAVES.

Each wave contains at any one moment particles in all possible stages of their oscillation; some rising, and some falling; some at the top, and some at the bottom; and the distance from any row of particles to the next row that is in precisely the same stage of oscillation is called *breadth* of the wave, that is, the distance from crest to crest, or from hollow to hollow.

There is a striking similarity between the rising and falling of waves and the vibrations of a pendulum, and it is a very interesting and remarkable fact, that a wave always travels its own breadth in precisely the same time that a pendulum, whose length is equal to that breadth, performs one vibration. Thus, a pendulum $39\frac{1}{8}$ inches long beats once in each second, and a wave whose breadth is $39\frac{1}{8}$ inches travels that breadth in one second. The length of a pendulum must be increased as the square of the time for its vibrations; that is, to beat but once in two seconds, it must be four times as long as for one second; to beat once in three seconds, it must be nine times as long, and so on. In the same way, waves which travel their breadth in two seconds are four times as wide as those traveling their breadth in one second; and thus their breadth, and consequently their speed, increases as the square of the time. Large waves, therefore, roll onward with far greater velocity than small ones. If only thirty-nine inches wide, they move about two and a quarter miles an hour, and pass once each second; if

13 feet wide, they move $4\frac{1}{2}$ miles an hour, passing once in 2 seconds.
52 do.   do.   9   do.   do.   4 do.
209 do.   do.   18   do.   do.   8 do.
836 do.   do.   36   do.   do.   16 do.

Although the water itself does not advance where there is much depth, yet when it reaches a shore or beach, the hard and shallow bottom prevents it from falling or subsiding, and it then rolls onward with a real progressive motion from the momentum it has acquired, breaks into foam, and lashes the earth and rocks. The sea billows are sometimes twenty-five feet in elevation,* and when these advance upon a stranded ship on a lee shore, with

---

* No authentic measurement gives the perpendicular height of waves more than twenty-five feet.

the speed of a locomotive, their effects are in the highest degree appalling; and iron bolts are snapped, and massive timbers crushed beneath their violence.

PREVENTING THE INROAD OF WAVES.

To prevent the inroads of lake waves upon land, the remedies must vary with circumstances. The difficulty would be small if the water always stood at the same height. The greatest mischief is usually done when they *rise over* the beach of sand and gravel which they have beaten for centuries. Wooden bulwarks soon decay. Where loose stones can be had in large quantities, forming sloping *rip-rap* walls, they may be cheapest; but they are not unfrequently placed too near low-water mark to protect the banks. Substances which offer a gradual impediment to the waves are often quite effectual, though not formidable in themselves. It is curious to observe how so slender a plant as the bulrush, growing in water several feet deep, will destroy the force of waves. If it grew only near the shore, where the water has progressive motion, it would soon be dashed in heaps on the beach. Parallel hedge-rows of the osier willow, protected by a wooden barrier until well grown and established would, in many cases, prove efficient.

Stones and timber bulwarks are often made needlessly liable to injury by being built nearly perpendicular, and the waves break suddenly, and with full force, like the blows of a sledge against them. A better form is shown in fig. 261, where a slope is first presented, to weaken their force without imposing a full resistance, and their strength is gradually spent as they rise in a curve. A

Fig. 261.

more gradual slope than the figure represents would be still better. It is on this principle that the stability of the world-renowned Eddystone light-house depends. The base spreads out in every direction, like the trunk of a tree at the roots; and although the spray is sometimes dashed over its lofty summit by the violence of the storm, it has stood unshaken on its rocky base far out in the sea, against the billows and tempests, for nearly a century.

An instance occurred many years ago in England, where the superiority of knowledge over power and capital without it was strongly exemplified. The sea was making enormous breaches on the Norfolk and Suffolk coast, and inundated thousands of acres. The government commissioners endeavored to keep it out by strong walls of masonry and breakwaters of timber, built at great expense; but they were swept away by the fury of the billows as fast as they were erected. A skillful engineer visited the place, and, with much difficulty, persuaded them to adopt his simple plan. Observing the slope of the beach on a neighboring shore, he directed that successive rows of fagots or brush be deposited for retaining the sand, which was carted from the hills, forming an embankment with a slope similar to that of the natural beach. Up this slope the waves rolled, and became gradually spent as they ascended, till they entirely died away. The breach was effectually stopped, and this simple structure has ever since resisted the most violent storms of the German Ocean.

CONTENTS OF CISTERNS.

Connected with the subject of hydraulics is the collection and security of water falling upon roofs, in all cases where a deficiency is felt by farmers in the drought of summer. The amount which falls upon most farm-buildings is sufficient to furnish a plentiful supply to all the domestic animals of the farm when other supplies fail, if cisterns large enough to hold it were only provided. Generally speaking, none at all are connected with barns and out-buildings, and even when they are furnished, they are usually so small as to allow four-fifths of the water to waste.

If all the rain that descends in the Northern States of the Union should remain upon the surface, without sinking in or running off, it would form, each year, a depth of about three feet. Every inch that falls upon a roof yields two barrels for each space ten feet square; and seventy-two barrels a year are yielded by three feet of rain. A barn thirty by forty feet supplies annually from its roof eight hundred and sixty-four barrels, or enough for more than two barrels a day for every day in the year. Many farmers have in all five times this amount of roof, or enough for twelve barrels a day, yearly. If, however, this water were collected, and kept for the dry season only, twenty or thirty barrels daily might be used.

In order to prevent a waste of water on the one hand, and to avoid the unnecessary expense of too large cisterns, their contents should be determined beforehand by calculation.

RULE FOR DETERMINING THE CONTENTS.

A simple rule to determine the contents of a cistern, circular in form, and of equal size at top and bottom, is the following :—Find the depth and diameter in inches; square the diameter, and multiply the square by the decimal .0034, which will give the quantity in gallons* for one inch in depth. Multiply this by the depth, and divide by

---

* This is the standard gallon of 231 cubic inches. The gallon of the State of New York contains 221.184 cubic inches, or 6 pounds at its maximum density.

31½, and the result will be the number of barrels the cistern will hold.

For each foot in depth, the number of barrels answering to the different diameters are,

For 5 feet diameter	4.66	barrels.
6 "	6.71	"
7 "	9.13	"
8 "	11.93	"
9 "	15.10	"
10 "	18.65	"

By the rule above given, the contents of barn-yard cisterns and manure tanks may be easily calculated for any size whatever.

The size of cisterns should vary according to their intended use. If they are to furnish a daily supply of water, they need not be so large as for keeping supplies for summer only. The average depth of rain which falls in this latitude, although varying considerably with season and locality, rarely exceeds seven inches for two months. The size of the cistern, therefore, in daily use, need never exceed that of a body of water on the whole roof of the building, seven inches deep. To ascertain the amount of this, multiply the length by the breadth of the building, reduce this to inches, divide the product by 231, and the quotient will be gallons for each inch of depth. Multiplying by 7 will give the full amount for two months' rain falling upon the roof. Divide by 31½, and the quotient will be barrels. This will be about fourteen barrels for every surface of roof ten feet square when measured horizontally. Therefore, a cistern for a barn 30 by 40 feet should hold one hundred and sixty-eight barrels; that is, as large as one ten feet in diameter, and nine feet deep. Such a cistern would supply, with only thirty inches of rain yearly, no less than six hundred and thirty barrels, or nearly two a day.

Cisterns intended only for drawing from in times of drought, to hold all the water that may fall, should have about three times the preceding capacity.

PART III.

MACHINERY IN CONNECTION WITH AIR.

CHAPTER I.

PRESSURE OF AIR.

PNEUMATICS treats of the mechanical properties of the air.

The actual weight of the air may be correctly found by weighing a strong glass vessel furnished with a stop-cock, a (fig. 262), after the air has been withdrawn from it by means of an air-pump. Let it be accurately balanced by weights in the opposite scale; then turn the stop-cock and admit the air, and it will immediately descend, as shown in the figure. The weight of the admitted air may be ascertained by adding weights until it is again balanced.

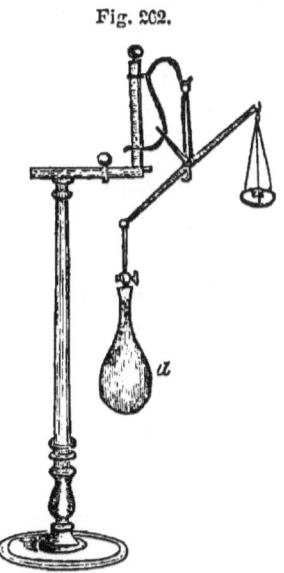

Fig. 262.

Balance for Weighing Air.

HEIGHT AND WEIGHT OF THE ATMOSPHERE.

The atmosphere which covers the earth extends upward to a height of about fifty or sixty miles. At the surface of the earth the air is about eight hundred times lighter than the water, and the higher we ascend, the rarer or lighter it becomes, from the diminished pressure of the weight above. At seven miles high, it is four times lighter than at the surface; at twenty-one miles, it is sixty-four times lighter; and at fifty miles, about twenty thousand times lighter. At this height it ceases to refract the rays of the sun so as to render it visible at the earth's surface; but if it decreases at the same rate upward, at a hundred miles high it must be nearly a thousand million times rarer than at the earth.

If the atmosphere were uniformly of the same density, with its present weight, it would reach only five miles high. Although so much lighter than water, yet, from its great height, it presses upon the surface of the earth as heavily as a depth of thirty-three feet of water. This is nearly equal to fifteen pounds on every square inch, or more than two thousand pounds to the square foot. This enormous weight would instantly crush us, did not air, like liquids, press in every direction, so that the upward exactly counterbalances the downward pressure, and the air within the body counteracts that without.

The weight of the atmosphere is strikingly shown by means of an air-pump, which pumps the air from a glass vessel, placed mouth downward upon the brass plate of the machine (fig. 263). When the air is pumped out, and the upward or counterbalancing air removed, so heavy is the load upon the glass vessel, that a strong man could scarcely remove it from the plate, although it

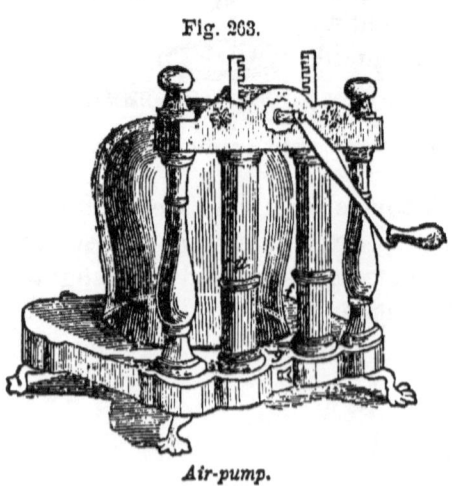

Fig. 263.

Air-pump.

WEIGHT OF THE ATMOSPHERE.

be no larger than a small tumbler. A glass jar with a mouth six inches across would need a force equal to nearly four hundred pounds to displace it. If there be a glass vessel open at both ends, the hand placed on the top may be so firmly held by the pressure that it can not be removed until the air is again admitted below (fig. 264). If a thin plate of glass be placed on the top of this open vessel, on pumping out the air, the weight will suddenly crush it with a noise like the report of a gun.

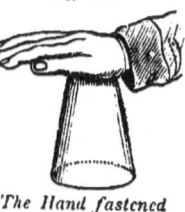

Fig. 264.

The Hand fastened by Air.

Some interesting instances occur in nature of the use of atmospheric pressure. Flies walk on glass by means of the pressure against the outside of their feet, the air having been forced out beneath. In a similar way, some kinds of fishes cling to the sides of rocks under water, so as not to be swept off by the current. Dr. Shaw threw a fish of this kind into a pail of water, and it fixed itself so firmly to the bottom, that, by taking hold of the tail, he lifted up the pail, water and all.

It is the pressure of the atmosphere upon water that drives it up the barrel of a pump as soon as the air is pumped out from the inside. Hence the reason that pumps can never be made to draw water more than thirty-three feet below the piston, a height corresponding to the weight of the atmosphere. In practice they never draw water even to this height, as a perfect vacuum can not be made by pumping.

THE BAROMETER.

On the same principle the *Barometer* is made. It consists of a glass tube, nearly three feet long, open at one end, and which is first filled with mercury, a liquid nearly fourteen times heavier than water. The open end is then placed downward in a cup of mercury. The weight of the mercury in the tube causes it to descend until the pressure of the atmosphere on the mercury in the cup preserves an equilibrium, which takes place when the column in the tube has fallen to about two feet and a half high, the upper part of the tube being left a perfect vacuum, as no air can enter (fig. 265). Now, as the height of the column of mercury depends alone upon the weight of the atmosphere, then, whenever the air becomes lighter or heavier, as it constantly does during the changes of the weather, the rising or falling of the column indicates these changes; and, what is very important, it shows the approaching changes of the weather several hours before they actually take place. Hence it becomes a valuable assistant in foretelling the weather. When the mercury falls, showing that the atmosphere is becoming lighter, it indicates the approach of storms or rain; when it rises, a settled or fair sky follows. These are often foreshown before there is any change in the appearance of the sky. For this reason the barometer is sometimes called a *weather-glass*. It is of the greatest value to navigators at sea. Long voyages, which formerly required a year, have been made in eight months by means of the assistance afforded by the barometer, admitting a full spread of canvas by night as well as by day, from the certainty of its predictions. On land its indications are not so certain, and at some places less so than at others. Sometimes, and more commonly during autumn and winter, the sinking of the mercury is followed only by wind instead of rain. There is, however, no doubt that its use would be of much advantage in large farming establishments, more especially during the precarious seasons of haying and harvesting.

The barometer is an instrument of great value in determining with little labor, and with considerable accuracy,

Fig. 265.

Barometer.

the heights of mountains, hills, and the leading points of an extensive district of country. In rising above the level of the sea, the weight of the air above us becomes less; that is, the pressure of the air upon the barometer decreases, and the column of mercury gradually falls as we ascend. To determine, therefore, the height of a mountain, we have only to place one barometer at its foot while another stands at the top, and then, by observing the difference in the height of the mercury, we are enabled to calculate the height of the mountain. The following table shows how much the barometer falls at different altitudes, thirty inches being taken for the sea-level:*

At 1,000 feet above the sea, the column falls to 28.91 inches.
    2,000   "   "   "   "   "   27.86 "
    3,000   "   "   "   "   "   26.85 "
    4,000   "   "   "   "   "   25.87 "
    5,000   "   "   "   "   "   24.93 "
      1 mile   "   "   "   "   24.67 "
      2 "   "   "   "   "   20.29 "
      3 "   "   "   "   "   16.68 "
      4 "   "   "   "   "   13.72 "
      5 "   "   "   "   "   11.28 "
     10 "   "   "   "   "   4.24 "
     15 "   "   "   "   "   1.60 "
     20 "   "   "   "   "   0.95 "

At the level of the sea, the barometer falls about one hundredth of an inch for a rise of nine feet, or a little more than the tenth of an inch for a rise of one hundred feet. At a height of one mile it requires about eleven feet rise to sink the mercury a hundredth of an inch.

In selecting land in mountainous districts of the country, where degrees of frost increase with increased altitudes, and where the height of one portion above another has an important relation to the cost of drawing loads up

---

* The mercury rarely stands as high as 30 inches at the level of the sea, the mean height being about 29.5 inches. But this does not affect the measurement of heights, which is determined, not by the actual height, but by the *difference* in heights.

and down hill, the barometer might become of much practical value.

THE SYPHON.

The *syphon* operates on a principle quite similar to that of the pump; but, instead of pumping out the air of the tube through which the water rises, a vacuum is created  by the weight of a column of water, in the following way: Fig. 266 represents a syphon, which is nothing more than a tube bent in the form of a letter U inverted. Now, if this be filled throughout with water, and then placed with the shorter arm in the vessel of water, A, the weight of the column of water in the longer arm, which is outside, will overbalance the weight of the other column, and will therefore run out in a stream. This tends to cause a vacuum in the tube, which is instantly filled by the water rushing up the shorter arm, being driven up by the pressure of the atmosphere. A stream will consequently continue running through the syphon until the vessel is drained.

Fig. 266.

The syphon may sometimes be very usefully employed in emptying pools or ponds of water on high ground, without the trouble of cutting a ditch for this purpose. For 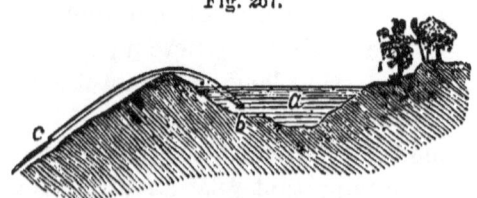 instance, let a (fig. 267) represent a body of water which it is desirable to drain off; by placing the lead tube, $b\ c$, so that the arm, c, may be lowest, and applying a pump at this arm to withdraw the air and fill the syphon with water, it will commence running, and continue until the

Fig. 267.

water has all been drawn off. Difficulties, however, sometimes occur. If the tube is small and very long, and the descent is trifling, the friction of the water in the tube may prevent success. Water usually gives out small quantities of air, which collects in the higher part of the syphon, and after a while fills it, causing the stream to cease running; but syphons for this purpose, when only a few rods in length, with several feet descent, are usually found to succeed well. If the discharging orifice is several times smaller than the tube, it is frequently of material use, by causing a slow and steady current through the syphon.

CHAPTER II.

MOTION OF AIR.

WINDS.

Wind is air in motion. Its force depends on its speed. When its motion is slow, it constitutes the soft, gentle breeze. As the velocity increases, the force becomes greater, and the strong gale sweeps around the arms of the wind-mill with the strength of many horses, and huge ships are driven swiftly through the waves by its pressure. By a still greater velocity of the air, its power becomes more irresistible, and solid buildings totter, and forest trees are torn up by the roots in the track of the tornado.

The force of wind increases directly as the square of the velocity. Thus a wind blowing ten miles an hour exerts a pressure four times as great as at five miles an hour, and twenty-five times as great as at two miles an hour.

The following table exhibits the force of wind at different degrees of velocity:

Miles an hour.	Pressure in lbs. on a square foot.	Description.
1	.005	Hardly perceptible.
2	.020	Just perceptible.
3	.045	
4	.080	Light breeze.
5	.125	
6	.180	Gentle, pleasant wind.
7	.320	
10	.500	Pleasant, brisk wind.
15	1.125	
20	2.000	Very brisk.
25	3.125	
30	4.500	Strong, high wind.
35	6.125	
40	8.000	Very high.
45	10.125	
50	12.500	Storm or tempest.
60	18.000	Great storm.
80	32.000	Hurricane.
100	50.000	Tornado, tearing up trees, and sweeping off buildings.

These forces may be observed at a time when the air is still, by a forward motion equal to that of the wind. Thus walking moderately gives the faint breeze against the face; riding in a wagon at six miles an hour causes the sensation of a pleasant wind; the deck of a steam-boat at fifteen miles produces a brisk blow; while an open rail-car at forty miles an hour occasions a sweep of the air nearly resembling a tempest.

The preceding table will enable any one to calculate with considerable accuracy the amount of draught which a horse must constantly overcome in traveling with a covered carriage against the wind, adding, of course, the speed of the horse to that of the wind. For example, suppose a horse with a covered carriage is driven against what we term "a very brisk wind," blowing 24 miles an hour, and pressing 3 lbs. on the square foot. The carriage top offers a resisting surface four feet square, or with sixteen square feet. Three times sixteen, or 48 lbs., are consequently required to be overcome with every onward step of the horse. Now, we have already seen, when treating of "application of labor," that a horse traveling three miles an hour for eight hours will overcome only 83 lbs. with ordinary working, which is not double the resistance of the wind. Hence we perceive that more than half the horse's strength is lost by driving against such a current. At six miles an hour, all his strength, without over-driving, would be expended in overcoming the force of the wind, and the power required for moving the carriage would be so much excess of labor. For simplifying the operation, the increased motion of the wind occasioned by driving against it has not been taken into account.

Even with a small pressure, the loss in power is considerable for an entire day. When, for example, the air is perfectly still, traveling six miles an hour will cause a constant resistance of 3 lbs. on the carriage, or one-fourteenth of the power exerted for a full day's work. The same speed against a "gentle wind" of six miles an hour, added, would increase the resistance fourfold, or equal to 12 lbs.; more than one-fourth of the horse's strength at six miles an hour through the day.

WIND-MILLS.

The power possessed by the sails of a wind-mill may be nearly ascertained in the same way, the area of the sails being known, and first deducting their average velocity.

In all wind-mills, it is important that the sails should have the right degree of inclination to the direction of the wind. If they were to remain motionless, the angle would be different from that in practice. They should more nearly *face* the wind; and as the ends of the sails sweep around through a greater distance and faster, they should present a flatter surface than the parts nearer the centre. The sails should, therefore, have a twist, to give them the most perfect form, so that the parts nearest the centre may form an angle of about 68 degrees with the wind, the middle about 72 degrees, and the tips about 83 degrees.

In order to produce the greatest effect, it is necessary to give the sails a proper velocity as compared with the velocity of the wind. If they were entirely unloaded, the extremities would move faster than the wind, in consequence of its action on the other parts. The most useful effect is produced when the ends move about as fast as the wind, or about two-thirds the velocity of the average surface.

The most useful wind is one that moves at the rate of eight to twenty miles per hour, or with an average pressure of about one pound on a square foot. In large windmills, the sails must be lessened when the wind is stronger than this, to prevent the arms from being broken; and if much stronger, it is unsafe to spread any, or to run them.

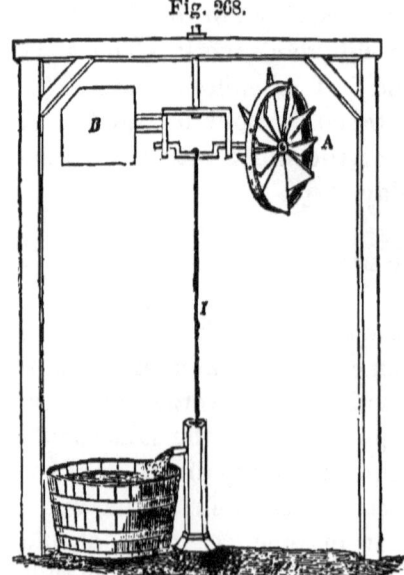

Fig. 208.

Wind-mill for pumping water on farms: A, wind-mill; B, vane; I, pump-rod.

The force of wind may be usefully applied by almost every farmer, as it is a universal agent, possessing in this respect great advantages over water-power, of which very few farms enjoy the privilege.

Wind may be applied to various purposes, such as sawing wood by the aid of a circular saw, turning grindstones, and particularly in pumping water. One of the simplest contrivances for pumping is represented by fig. 268, where A is the circular wind-mill, with a number of sails set obliquely to the direction of the wind, and always kept facing it by means of the vane, B. The crank of the wind-mill, during its revolutions, works the pump-rod, I, and raises the water from the well beneath. In whatever di-

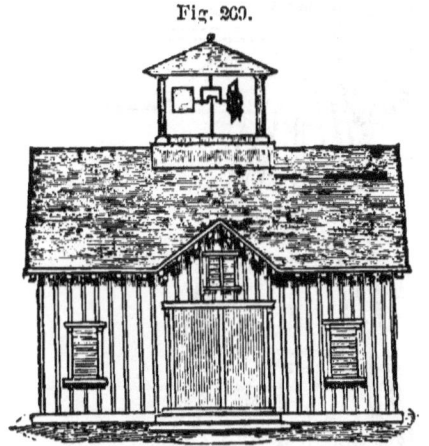

Fig. 269.

Barn surmounted with wind-mill for pumping water, cutting straw, &c.

rection the wind may blow, the pump will continue working. The pump-rod, to work steadily, must be immediately under the iron rod on which the vane turns. If the diameter of the wind-mill is four feet, it will set the pump in motion even with a light breeze, and with a brisk wind will perform the labor of a man. Such a machine will pump the water needed by a herd of cattle, and it may be placed on the top of a barn, with a covering, to which may be given the architectural effect of a tower or cupola, as shown in fig. 269.

A more compact machine, but of more complex construction, is shown in fig. 270, where the upper circle moves around with the wheel and vane on the fixed lower circle, to which it is strongly secured so as to admit of turning freely. In other respects it is similar to the preceding.

11*

250   MACHINERY IN CONNECTION WITH AIR.

Fig. 270.

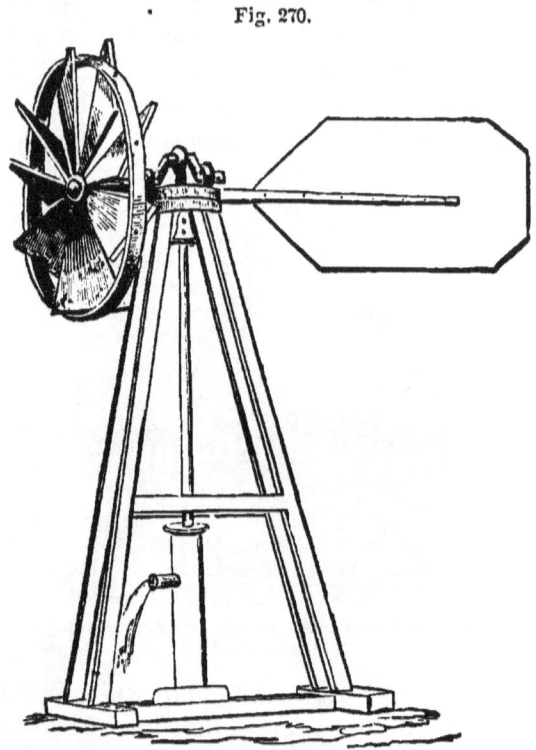

Wind-mills, like the preceding, which have fixed sails, *should not be more than three or four feet in diameter*, and even then will require care in storms. If larger, they will become broken by severe winds. The remedy is either to move the sails by hand at every considerable change in the force of the current, which would require nearly constant attention; or to use the self-regulating machines, of which there have been several invented, some of which have proved useful and durable.

Halliday's wind-mill has been much used for several years, and is made of various sizes, the larger possessing the power of several horses. It is self-regulated, in the following manner: When the mill begins to run too fast, it pumps water rapidly into a chamber or cylinder, and this increase of water moves an arm which turns the fans edgewise to the wind. When the wind slackens, a reverse movement takes place.

Brown's wind-mill, made by the Empire Wind-mill Company, of Syracuse, is a more recent invention, and has proved very successful. The annexed figure (fig. 271), rep-

Fig. 271.

Brown's (or Empire) Wind-mill.

resents one of the smaller sizes, adapted to farm purposes and pumping water for cattle. It is regulated in part by the centrifugal force of weights, and partly by the direct pressure of the wind. This regulating contrivance renders the mill safe, even in a gale of wind. The larger sizes, which are fifty feet or more in diameter, possess much power, and are used for grinding grain, and other purposes.

The work which a wind-mill is capable of doing depends very much on the site. If placed where the wind has a long, uniform, and steady sweep, it will accomplish much more, and to better satisfaction, than if among hills or other obstructions, where the blasts are uncertain and changing.

Wind-mills of large size are peculiarly adapted to pumping water into reservoirs, or from mines or quarries, where a few days of calm weather will not result in inconvenience; but they are not suited to manufactories where a constant power is required to furnish employment to men, but can be used for work which may be intermitted or changed.

Brown's wind-mill is sold at $75 dollars for the small size, with increase of prices up to $1,200 for large ones.

CAUSES OF WIND.

The motion of air, in producing wind, is explained by the action of heat, *although there are many irregular currents whose cause is not well understood.* The simplest illustration of the effect of heat in causing currents is furnished by the land and sea-breezes in warm latitudes. The rays of the sun during the day heat the surface of the land, and the air in contact with it, also becoming heated, and thus rendered lighter, flows upward; the air from the sea rushes in to fill the vacancy, and causes the sea-breeze. During the night, the radiation of heat from the land into the clear sky above cools the surface to a lower temperature than that of the sea; consequently, the air in contact with the sea becomes heated the most, and rising, causes the wind from the land to flow in and supply the place. Trade-winds are caused in a similar way, but on a much larger scale, by the greater heat of the earth at the equator, which produces currents from colder latitudes. These currents assume a westerly tendency, in consequence of the velocity of the earth being the greatest at the equator, and which, outstripping the momentum which the winds have acquired in other latitudes, tends to throw them behind, or in a westerly direction.

CHIMNEY CURRENTS.

Chimney Currents are produced by the heat of the fire rarefying the air, which rises, and carries the smoke with it. The taller the chimney is, the longer will be the column of rarefied air tending upward; and, as a consequence, the stronger will be the draught. In kindling a fire in a cold chimney, there is very little current till this column becomes heated. The upward motion of heated currents is governed by laws similar to the downward motion of water in tubes, where the velocity is increased with the height of the head. But as air is more than eight hundred times lighter than water, slight causes will affect its currents, which would have no sensible influence on the motion of liquids. For instance, a strong wind striking the top of a chimney may send the smoke downward into the room; and a current can not be induced through a horizontal pipe without connecting with it an upright pipe of considerable height.

CONSTRUCTION OF CHIMNEYS.

In constructing chimneys to produce a strong draught, the throat immediately above the fire, which should have a breadth equal to that of the fireplace, should be contracted to a width of about four inches, so that the column of rising air above may draw the air up through the throat with increased velocity, as shown in fig. 272. This arrangement also allows the fire to be built so as to throw the heat more fully out into the room. By leaving the shoulder at *b* square or flat, it will tend to arrest any reversed or downward current in a better manner than if built sloping, as shown by the dotted line at *a*, which would act like a funnel, and throw the smoke into the room. The throat should be about as high as the extreme tip of the flame; if much higher, the chimney will not draw so well, and if lower, too much of the heat will be lost.

Fig. 272. *A well-built Chimney.*

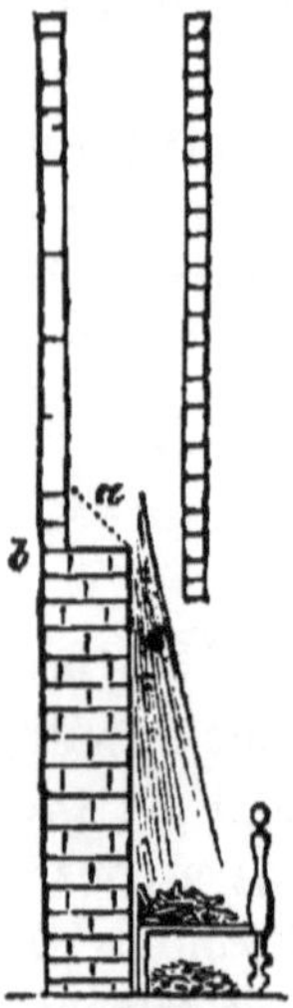

Fig. 273. *A badly-built Chimney.*

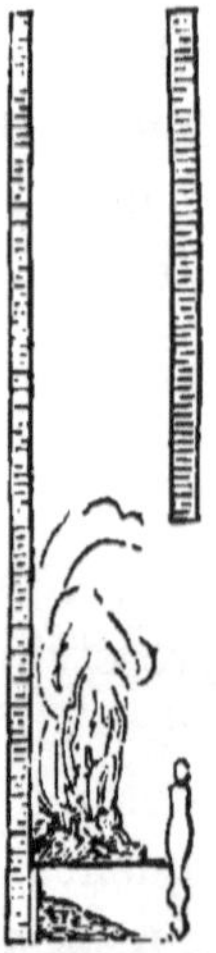

Fig. 273 shows a fireplace without a contracted throat, the current of which is comparatively feeble. Many chimneys draw badly by being made too large for the fire to heat sufficiently the column of air they contain.

CHIMNEY-CAPS.

When wind sweeps over the roof of a high part of the building, or over a hill, it often strikes the tops of chimneys below, and drives the smoke downward. This may be often prevented by placing a cap over the chimney, like that represented by fig. 274, which is supported at its corners, the smoke passing out at the four sides just under the eaves of this cap. But it sometimes happens that there is a confusion of currents and eddies at the top of the chimney, over which this cap

Fig. 274.

CHIMNEY-CAPS. 255

has no influence. In this case, the cap represented by fig. 275 furnishes a complete remedy, and is, indeed, perfect in its operation under any circumstances whatever, for the chimney surmounted by it will always draw when there is wind from any quarter, with or without a fire. It has effected a perfect cure in some chimneys which before were exceedingly troublesome, and were regarded as incurable. Fig. 276 is intended to show the mode of its operation; the wind, as shown by the arrows, being deflected for a considerable distance on the lee side, so as to form a vacancy at a, which the wind from the other end and from the chimney both rush in to supply. Being fixed on without turning in the chimney, it is both simpler and less noisy than any caps furnished with a vane.

Fig. 275.

Fig. 276.

Emerson's Chimney-cap is different in construction, but quite similar in principle to the preceding. It is shown by fig. 277. A sheet-iron pipe is set in the top of the chimney, furnished with the conical rim, and a plate or fender on the top, which excludes the rain. Between the plate and rim is a space quite similar in form or section to that represented by fig. 276.

Fig. 277.

Fig. 278.
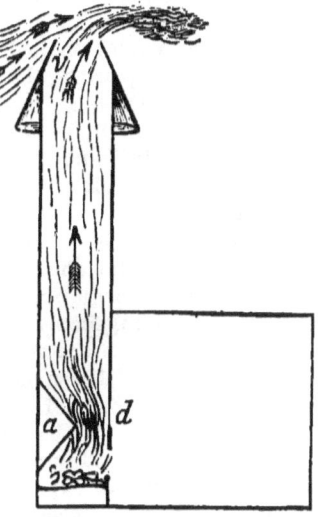

In exposed situations, chimneys are found to draw more uniformly by contracting

Fig. 279.

the top about a third less than the rest of the flue. The current at the moment of escape is swifter than below, and less acted upon by any downward blast of the wind, at the same time that the surface is smaller on which the wind can strike the current, as shown in fig. 278. A chimney of this character may be very easily made by contracting the tiers of brick, thus giving to it an ornamental appearance, as seen in fig. 279.*

---

* Where different fires communicate with the same chimney, separate flues should be built for each fire, and kept separate in the same chimney-stack, carried up independently of each other. But even with this precaution, smoky rooms will not be avoided, unless the termination of the chimney is of the right form, of which the following illustration is given in Allen's Rural Architecture:

"Fifteen years ago we purchased and removed into a most substantial and well-built stone house, the chimneys of which were constructed with open fireplaces, and the flues carried up separately to the top, where they all met upon the same level surface, as chimneys in past times usually were built thus (fig. 280). Every fireplace in the house (and some of them had stoves in) smoked intolerably; so much so, that when the wind was in some quarters, the fires had to be put out in every room but the kitchen, which, as good luck would have it, smoked less—although it did smoke there—than the others. After balancing the matter in our own mind some time whether we would pull down and rebuild the chimneys altogether, or attempt an alteration—as we had given but little thought to the subject of chimney draught, and to try an experiment was the cheapest— we set to work a bricklayer, who, under our direction, simply built over each discharge of the several flues a separate top of fifteen inches high, in this wise: Fig. 281. The remedy was perfect. We have had no smoke in the house since, blow the wind as it may, on any and on all occasions. The chimneys *can't* smoke; and the whole expense for four chimneys, with their twelve flues, was not twenty dollars! The remedy was in giving each outlet a *distinct* current of air all around, and on every side of it."

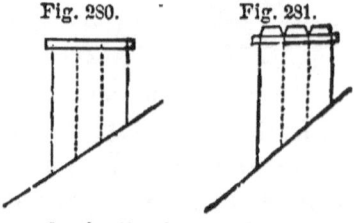

Fig. 280.   Fig. 281.

VENTILATION.

Impure air may be breathed for a short time without any serious detriment, but to live in it and respire it for years can not fail to produce permanent injury to the health. During the heat of summer, open doors and windows will usually furnish plenty of fresh air, so long as this season lasts, which in the Northern States is not one half the year. During the rest of the time, rooms are heated with close stoves, and unless special care is taken to secure fresh air, pale or sickly inmates will be the most likely results.

Even with a common open fireplace, which causes more circulation of the air in a room than stoves, the ventilation is very imperfect. The following figure (fig. 282) represents the fresh air as passing in from an open window opposite the fire, producing a direct current from the window to the chimney, and leaving all the upper portion of the room filled with bad air, unaffected by the change. The cold air can not rise, nor the hot air descend. This difficulty may be easily removed by placing a register (which may be closed or opened at pleasure) at a, in the upper corner, so that the confined air may escape into the chimney. Without this provision, it is nearly impossible to preserve the air in proper condition for breathing, for the upper part, being warmest and lightest, remains unchanged at the top. In rooms heated by stoves, registers for the escape of the foul air are still more important, where

Fig. 282.

A badly-ventilated Room.

the thermometer frequently indicates twenty degrees difference in the heat above and at the floor, the lower stratum of air resting like a cold lake about the feet, while the head is heated unduly.

When the draught of the chimney-fire is not strong, the smoke may, however, escape through the ventilating register into the room. To avoid this difficulty, it is best to provide separate air-flues in the walls when the house is built, for effecting perfect ventilation. In rooms strongly heated by fires, the fresh air should be admitted near the ceiling, producing descending currents, and effecting a complete circulation in the air of the room. But in sleeping apartments, and in closets, not heated artificially, and where the descending currents will not take place, the fresh air should be admitted through a register or small rolling blind near the floor, and discharged near the ceiling into an air-flue.

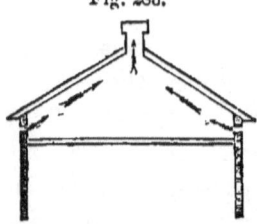

Mode of Ventilating Garrets.

The excessive warmth of garrets in midsummer may

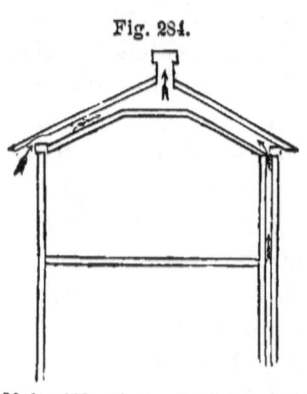

Mode of Ventilating half-story Bed-rooms.

Griffith's Ventilator.

be avoided by placing a ventilator at the highest part,

and admitting air at windows or openings near the eaves (fig. 283), thus sweeping all the hot air out by the current produced; or the oppressive heat of half-story bedrooms may be similarly avoided, by creating a current of air between the roof and the plastering (fig. 284). Two modes may be adopted, as represented on each side of the figure.

Fig. 285 represents Griffith's patent ventilator, for chimneys, and for giving a current of air through apartments. It is made of iron, working as a *screw fan*, the slightest wind causing it to revolve and establish a current through the pipe which it surmounts.

PART IV.

HEAT.

CHAPTER I.

CONDUCTING POWER OF BODIES.

When any substance or body has become heated, it loses its heat in two different ways, by *conduction* and by *radiation*. When conducted, heat passes off slowly or gradually through bodies, as when a pin is held by the hand in a candle, the heat advances from one end to the other till it burns the fingers; or, when an iron poker is thrust into the fire, the heat gradually passes through it till the whole becomes hot. Iron and brass are, therefore, said to be good conductors of heat. The end of a pipe-stem may, however, be heated to redness, and a wooden rod may be set on fire without even warming the other extremity, because the heat is very slowly conducted through them. Wood and burned clay are, therefore, *poor* conductors.

The comparative conducting power of different substances may be shown by placing short rods of each with one of their ends in a vessel of hot sand, the others to be tipped with wax. The different periods of time required to melt the wax indicate the relative conducting powers. It will speedily melt on the copper rod; soon after, on the rod of iron; glass will require longer time; stone or earthenware, still longer; while on a rod of wood, it will scarcely melt at all. These rods should be laid horizontally, that the hot air rising from the sand may not affect the wax. The conducting powers may be judged of, likewise, with considerable accuracy in cold weather, by merely placing the hand upon the different substances. The best conductors will feel coldest, because they withdraw the heat most rapidly from the hand. Iron will feel colder than stone; stone colder than brick; wood, still less so; and feathers and down, least of all, although the real temperature of all may be precisely the same.

UTILITY OF THIS PRINCIPLE.

A knowledge of this property is often very useful. For instance, it is found that hard and compact kinds of wood, as beach, maple, and ebony, conduct heat nearly twice as rapidly as light and porous sorts, like pine and bass-wood. Hence, doors and partitions made of light wood make a warmer house than those that are more heavy and compact. Pine or bass-wood would, in this respect, be better than oak or ash.

Porous substances of all kinds are the poorest conductors; sawdust, for example, being much less so than the wood that produced it. For this reason, sawdust has been used as a coating around the boilers of locomotives, to keep in the heat, and for the walls of ice-houses, to exclude it. Sand, filled in between the double walls of a dwelling, renders it much warmer in winter, and cooler in summer, than if *sandstone* were made to fill the same space. Ashes, being more porous, are found to be still better. Tan, which is similar to sawdust, is well adapted to filling in the walls of stables and poultry-houses, where more than usual warmth in winter is required. Confined air is a very poor conductor of heat; hence the advantage of double walls and double windows, provided there are no crevices for the escape of the confined air. This principle has been lately applied in the manufacture of *hollow brick* for building the walls of dwellings.

The light and porous nature of snow renders it eminently serviceable as a clothing to the earth in the depth of winter, preventing the escape of the heat from below, and protecting the roots of plants from injury or destruction. Hence the very severity of the cold of the Northern regions, by producing an abundance of those beautiful feathery crystals which form snow, becomes the means of protecting from its own effects the tender herbage buried beneath this ample shelter.

CONDUCTING POWER OF LIQUIDS.

Liquids are found to conduct heat very slowly, and they were for a long time considered perfect non-conductors. Some interesting experiments have been performed in illustration of this property. A large glass jar may be filled with water (fig. 286), in which may be fixed an air thermometer, which is always quickly sensitive to small quantities of heat. A shallow cup of ether, floating just above the bulb, may be set on fire, and will continue to burn for some time before any effect can be seen upon the thermometer. The upper surface of a vessel of water has been made to boil a long time with a piece of unmelted ice at the bottom. Liquids are found, however, to possess a conducting power in a very slight degree.

Fig. 287.

Fig. 286.

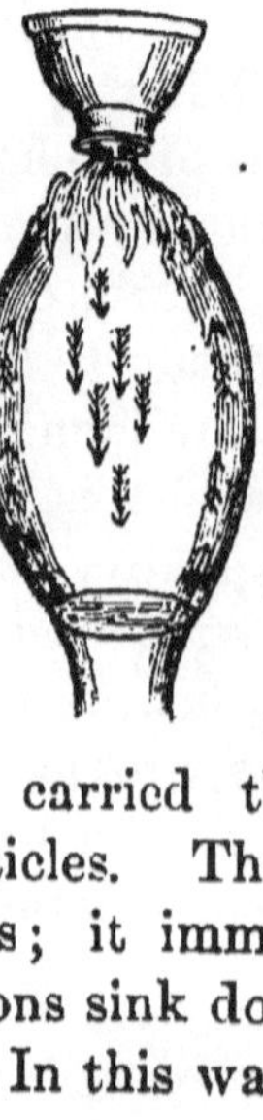

When a vessel of water is heated in the ordinary way over a fire, the heat is carried through it merely by the motion of its particles. The lower portion becomes warm, and expands; it immediately rises to the surface, and colder portions sink down and take its place, to ascend in their turn. In this way, a constant circulation is kept up among the particles. These rising and descending currents are shown by the arrows in fig. 287. This result may be easily shown by filling a flask with water into which a quantity of sawdust from some green hard wood has been thrown, which is about as heavy as water. It will traverse the vessel in a manner precisely as shown in the figure.

These results indicate the importance of applying heat directly to the bottom of all vessels in which water is intended to be heated. A considerable loss of heat often occurs when the flame is made to strike against the sides only of badly arranged boilers.

EXPANSION BY HEAT.

An important effect of heat is the expansion of bodies. Among many ways to show it, an iron rod may be so fitted that it will just enter a hole made for the purpose in a piece of sheet-iron. If the rod be now heated in the fire, it expands and becomes larger, and can not be thrust

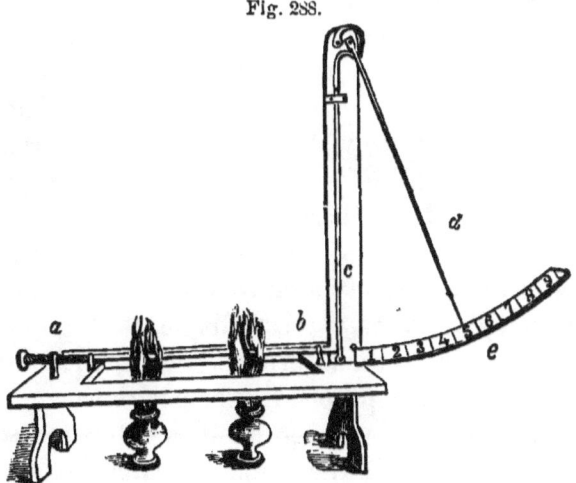

Fig. 288.

into the hole. The expansion may be more visibly shown and accurately measured by means of an instrument called the *Pyrometer* (fig. 288). The rod $a\ b$, secured to its place by a screw at *a*, presses against the lever *c*, and this against the lever or index *d*, both of which multiply the motion, and render the expansion very obvious to the eye when the rod is heated by the lamps. If the rod should expand one-fiftieth of an inch, and each lever multiplies twenty times, then the index (or second lever) will move along the scale eight inches; for 20 times 20 are 400, and 400-50ths of an inch are 8 inches.

Many cases showing the expansion of heated bodies occur in ordinary practice. One is afforded by the manner in which the parts of carriage wheels are bound together. The tire is made a little smaller than the wooden part of the wheel; it is then heated till, by expanding, it becomes large enough to be put on, when it is suddenly cooled with water, and, by its powerful contraction, binds every part of the wheel together with great force. Hogsheads are firmly hooped with iron bands in the same way, with more force than could ever be given by driving with blows of the mallet.

This principle was very ingeniously applied in drawing together two expanding brick walls of a large building in Paris, which threatened to burst and fall. Holes were drilled in the opposite walls, through which strong iron bars across the building projected, and circular plates of iron were screwed on these projecting ends. The bars were then heated, which increased their length; the plates were next screwed closely against the walls. On cooling, they contracted, and drew the walls nearer together. The process was repeated on alternating bars, until the walls were restored to their perpendicular positions.

All tools, where the wooden handles enter iron sockets, will hold more firmly if the metal is heated before inserting the wood.

The metallic parts of pumps sometimes become very difficult to unscrew, and a case has occurred where two strong men could not start the screws, until a bystander

EFFECTS OF SUDDEN EXPANSION. 265

suggested that the outer piece be heated, keeping the inner cool, when a force of less than ten pounds quickly separated them. In other cases, where the large iron nuts have been thoughtlessly screwed, while warmed with the hands, on the cold metallic axles of wood-sawing machines in winter, they have contracted so that the force of two or three men has been insufficient to turn them.

The sudden expansion of bodies by heat sometimes causes accidents. Thick glass vessels, when unequally heated, expand unequally, and break. Heated plates of cast-iron or cast kettles are liable to be fractured by suddenly pouring cold water upon them. The same effect has been usefully applied in splitting the scattered rocks which encumber a farm, and which are too large to remove while entire. Fires are built upon them; the upper surface expands while the lower remains cold, and large portions are successively separated in scales, and sometimes the whole rock is severed. The only care needed is to observe attentively and remove with an iron bar any parts which may have become loosened by the heat, and which would prevent the heat from passing to other portions. One man will thus attend to a large number of fires, and will split in pieces ten times as many rocks in a day as by drilling and blasting.

Fig. 289.

THE STEAM-ENGINE.

The *Steam-engine* owes its power to the enormous expansion of water at the moment it is converted into steam, which is about 1,600 times its bulk when in a liquid state. The principle on which the steam-engine acts may be understood by a simple instrument, represented in fig. 289. A glass tube with a small bulb is furnished with a solid, air-tight piston, capable of working up and

12

down. The water in the bulb, *a*, is heated with a spirit-lamp or sand-bath; the rising steam forces up the piston. Now, immerse the bulb in cold water or snow, and the steam is condensed again into water, the tube is left vacant, and the pressure of the atmosphere forces down the piston. By thus alternately applying heat and cold, it is driven up and down like the piston of a steam-engine. The only difference is, the steam-engine is furnished with apparatus so that this application of heat and cold is performed by the machine itself. The bulb represents the boiler, and the tube the cylinder; but in the steam-engine, the boiler is separate, and connected by a pipe with the cylinder; and instead of applying the cold water directly to the cylinder, it is thrown into another vessel, called the condenser, connected with the cylinder.

When Newcomen, who made the first rude regularly working engine, began to use it for pumping water, he employed a boy to turn a stop-cock connected with the condenser, every time the piston made a stroke. The boy, however, soon grew tired of this incessant labor, and endeavored to find some contrivance for relief. This he effected by attaching a rod from the piston or working-beam to the cock, which was turned by the machine itself at every stroke. This was the origin of the first self-acting engine.

The different parts of a common steam-engine may be understood from the following figures, one representing the boiler, and the other the working machinery.

The boiler, B (fig. 290), contains water in the lower part, and steam in the upper; FB is the fire; vo is the *feed-pipe;* v, a valve, closed by the lever bca, whenever the boiler is full enough, by means of the rising of the float, S, and opened whenever the float sinks from low water. M, *barometer gauge*, to show the pressure of the steam; w, weight on the lever, eb, for holding down the *safety-valve:* this lever being graduated like a steelyard,

THE STEAM ENGINE. 267

the force of the steam may be accurately weighed. *U* is a valve opening downward, to prevent the boiler being crushed by atmospheric pressure, by allowing the air to pass in whenever the steam happens to decline. Two

Fig. 290.

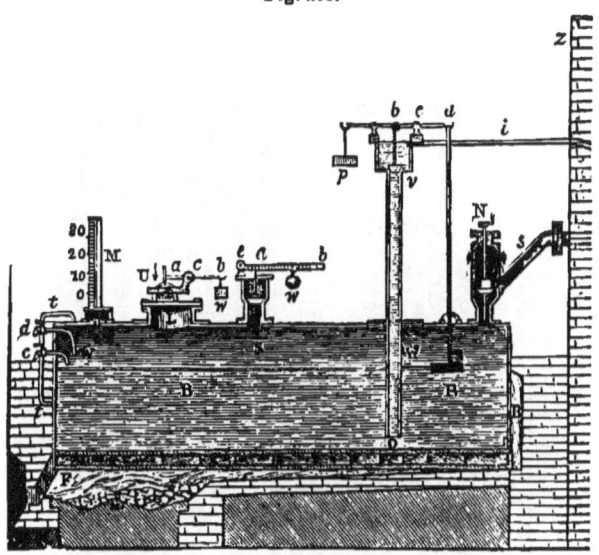

Boiler of Steam-engine.

tubes, with stop-cocks, *c* and *d*, one just below the water-level, and the other just above it, serve to show, by opening the cocks, whether the water is too high or too low.

The working part of the engine is represented in the figure on the following page (fig. 291). The steam enters by the pipe, *s*, from the boiler on the other side of the brick wall, as shown in fig. 290. The steam passes through what is called a *four-way-cock*, *a*, first into the lower, then into the upper end of the cylinder, *C*, as the piston, *P*, moves up and down; this is regulated by the levers, *y y*. The piston-rod, *E*, is attached to the working-beam, *B F*, turning on the centre, *A*. The rod, *F R*, turns the fly-wheel, *H H*, and drives the mill, steam-boat, or machinery to be set in motion.

The condenser, j, shown directly under the cylinder, remains to be described. It is immersed in a cistern of cold water, and is connected by pipes with the upper and lower end of the cylinder. Through these pipes the steam

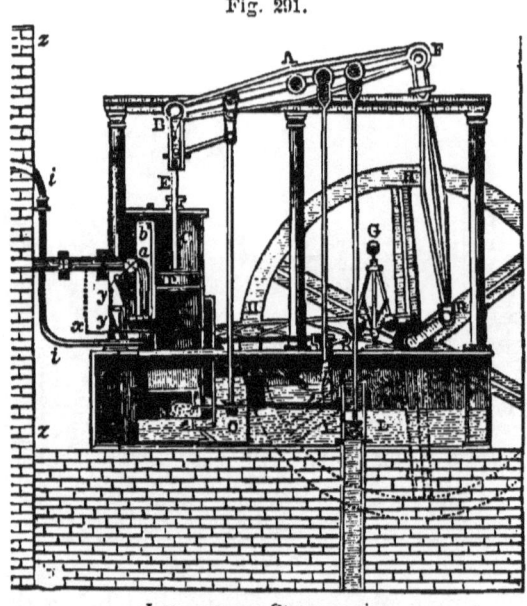

Fig. 201.

Low-pressure Steam-engine.

passes out of the cylinder, first from one end and then from the other, and is condensed into water by a jet of cold water thrown into it by the *injection-cock*. When condensed, it is pumped out by the pump, O, into the *well* or reservoir, W, and then again into the feed-pipe of the boiler. Warm water is thus constantly supplied to the boiler, and effects a great saving of fuel.

The supply of steam and the motion of the engine are regulated by the *governor*, G. When the motion is too fast, the two suspended balls, which revolve on a vertical or upright axis, and which hang loosely like pendulums, are thrown out from the axis, producing the movement of a rod which shuts the steam-valve. When the motion

QUALITIES OF THE STEAM-ENGINE. 269

is too slow, the balls approach the axis, and open the valve.

In *high-pressure* engines, the steam is not condensed, but escapes into the open air at every stroke of the piston, which produces the loud, successive *puffs* of all engines of this kind.

The steam-engine, in its most perfect form, is a striking example of human ingenuity, and its qualities are thus described by Dr. Arnott: "It regulates with perfect accuracy and uniformity the number of its strokes in a given time, and records them as a clock does the beats of its pendulum. It regulates the quantity of steam; the briskness of the fire; the supply of water to the boiler; the supply of coals to the fire. It opens and shuts its valves with absolute precision as to time and manner; it oils its joints; it takes out any air accidentally entering parts which should be vacuous; and when any thing goes wrong which it can not of itself rectify, it warns its attendants by ringing a bell; yet, with all these qualities, and even when exerting a force of six hundred horses, it is obedient to the hand of a child. Its aliment is coal, wood, and other combustibles. It consumes none while idle. It never tires, and wants no sleep. It is not subject to any malady when originally well made, and only refuses to work when worn out with age. It is equally active in all climates, and will do work of any kind; it is a water-pumper, a miner, a sailor, a cotton-spinner, a weaver, a blacksmith, a miller, a printer, and is indeed of all occupations; and a small engine in the character of a steam pony may be seen dragging after it, on an iron railway, a hundred tons of merchandise, or a thousand persons with the speed of the wind."

Steam-engines have been much used on large farms in England for thrashing, grinding the feed of animals, cutting fodder, and for other purposes. A successful English farmer has used a six-horse steam-engine to drive a pair of mill-stones, for thrashing and cleaning grain, elevating and bagging it, pumping water for cattle, cutting straw, turning a grindstone, and driving liquid manure through pipes for irrigating his fields, employing the waste steam in cooking food for cattle and swine. In this country, where horse labor is cheaper, steam-engines have not come into so general use; but on large farms, where a ten-horse-power or more is required, they have been employed to much advantage, consuming no food, and requiring no care

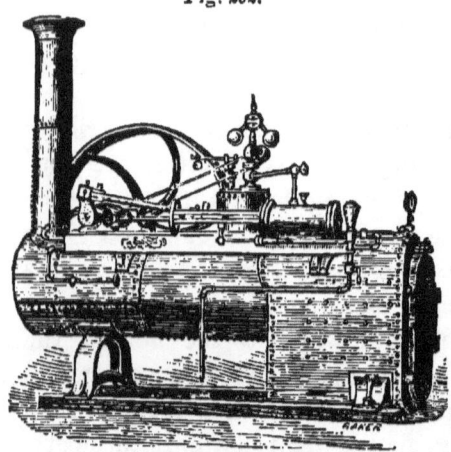

Fig. 292.
Wood's Farm Engine.

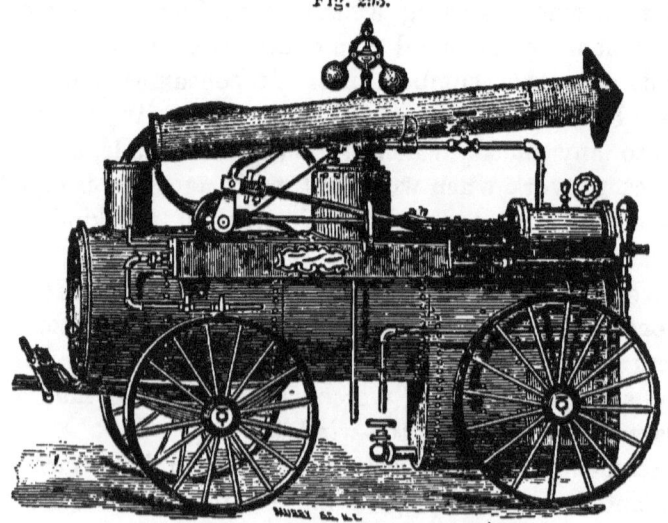

Fig. 293.
Wood's Engine on Wheels, with Pipe Folded Down.

when idle. Excellent steam-engines for this purpose are manufactured by A. N. Wood & Co., of Eaton,

Madison Co., N. Y., a representation of which is given in the accompanying figure (fig. 292.) When intended to move from place to place, these engines are furnished ready mounted on wheels (fig. 293). The twelve-horse-power engines cost about $1,000, and have thrashed over a hundred bushels per hour, using half a cord of wood, or 300 or 400 lbs. of coal for ten hours. A Western farmer thrashed 14,250 bushels of wheat in five consecutive weeks, working five and a half days each, with one of these engines. The smoke-pipe is guarded, so that straw placed within a few inches cannot be set on fire.

More difficulty obviously exists in adapting the steam-engine to plowing than for stationary purposes. In order to possess sufficient power, when used as a locomotive, the engine must be made so heavy as to sink in common soft soil even with large and broad wheels; and this tendency is increased by the jar of the machinery which these wheels support. For this reason, all locomotive plows have failed. Better success has attended the use of stationary engines, employed for drawing gangs of plows, by means of wire rope, across the fields. In England, where much of the soil is tenacious, and where fuel and manual labor are cheap, and horse labor expensive, this mode of plowing has been found profitable when employed on an extensive scale, and is now much used.

EXCEPTION TO EXPANSION BY HEAT.

A striking exception to the general law of expansion by heat occurs in the freezing of water.* During its change to a solid state, it increases in bulk about one-twelfth, and this expansion is accompanied with great force. The bottoms of barrels are burst out, and cast-iron kettles are split asunder, when water is suffered wholly to freeze in

---

* There are a very few other substances which expand on passing from a liquid to a solid state.

them. Lead pipes filled with ice expand; but if it is often repeated, they are cracked into fissures. A strong brass globe, the cavity of which was only one inch in diameter, was used by the Florentine academicians for the purpose of trying the expansive force of freezing water, by which it was burst, although the force required was calculated to be equal to *fourteen tons*. Experiments were tried at Quebec, in one of which an iron plug, nearly three pounds in weight, was thrown from a bomb-shell to the distance of 415 feet; and in another, the shell was burst by the freezing of the water which it contained.

This expansion has a most important influence in the pulverization of soils. The water which exists through all their minute portions, by conversion to frost, crowds the particles asunder, and when thawing takes place, the whole mass is more completely mellowed than could possibly be effected by the most perfect instrument. This mellowing is, however, of only short duration, if the ground has not been well drained to prevent its becoming again packed hard by soaking with water.

But this is not the most important result from the expansion of water. Much of the existing order of nature and of civilized life depends upon this property; without it the great mass of our lakes and rivers would become converted into solid ice; for, as soon as the surface became covered, it would sink to the bottom, beyond the reach of the summer's sun, and successive portions being thus added, the great body of all large rivers and lakes would become permanently frozen. But instead of this disastrous consequence, the ice, by resting upon the surface, forms an effectual screen from the cold winds to the water below.

LATENT HEAT.

If a vessel of snow, which has been cooled down to several degrees below freezing by exposure to the severe cold of winter, be placed over a steady fire with a thermometer in the snow, the mercury will rise by the increasing heat of the snow until it reaches the freezing point. At this moment it will stop rising, and the snow will begin to melt; and although the heat is all the time passing rapidly into the snow, the thermometer will remain perfectly stationary until it is all converted to water. The heat that goes to melt the snow does not make it any hotter; in other words, it becomes *latent* (the Latin word for *hidden*), so as neither to affect the sensation of the hand nor to raise the thermometer. Now it has been found that the time required to melt the snow is sufficient to heat the same quantity of water, placed over the same fire, up to 172 degrees, or 140 degrees above freezing; that is, 140 degrees have become latent, or hidden, in melting the snow.

This same amount of heat may be given out again by placing the vessel of water out of doors to freeze. A thermometer will show that the water is growing colder by the escape of the heat, until freezing commences. After this it still continues to pass off, but the water becomes no colder until all is frozen, as it was only the *latent* heat of the water that was escaping.

A simple and familiar experiment exhibits the same principle. Place a frozen apple, which thaws a little below freezing, in a vessel of ice-cold water. The latent heat of the water immediately passes into the apple and thaws it, and in an hour or two it will be found like a fresh apple and entirely free from frost; but the latent heat having escaped from the water next the apple, a thick crust of ice is found to encase it.

The amount of latent heat may be shown in still another way. Mix a pound of snow at 32 degrees, or at freezing, with a pound of water at 172 degrees. All will be melted, but the two pounds of water thus formed will be as cold as the snow, showing that for melting it the 140 degrees in the hot water were all made latent.

ADVANTAGES OF LATENT HEAT.

If no heat became latent by the conversion of ice and snow to water, no *time* would, of course, be required for the process, and thawing would be *instantaneous*. On the approach of warm weather, or at the very moment that the temperature of the air rose above freezing, snow and ice would all dissolve to water, and terrific floods and inundations would be the immediate consequence.

LATENT HEAT OF STEAM.

A still larger amount of latent heat is required for the conversion of water into steam; for, again place the vessel of water with its thermometer on the fire, it will rise, as the heat of the water increases, to 212 degrees, and then commence boiling. During all this time it will now remain stationary at 212, until the water is all boiled away. This is found to require nearly five times the period needed to heat from freezing to boiling; that is, nearly one thousand degrees of heat are made latent by the conversion of water into steam.

When the steam is condensed again to water, this heat is given out. Hence the use made of steam conveyed in pipes for heating buildings, and for boiling large vats or tubs of water, by setting free this large amount of latent heat which the fire has imparted to it.

GREEN AND DRY WOOD FOR FUEL.

A great loss is often sustained in burning green wood for fuel, from an ignorance of the vast amount of latent heat consumed to drive off the water the wood contains. When perfectly green, it loses about one-third of its weight by thorough seasoning, which is equal to about 25 cubic feet in every compact cord, or 156 imperial gallons. Now all this water must be evaporated before the wood is burned. The heat thus made latent and lost, being five times as great as to heat the water to boiling, is equal to enough for boiling 780 imperial gallons in burning up every cord of green wood. The farmer, therefore, who burns 25 green cords in a winter, loses heat enough to boil more than fifteen thousand gallons of water, which would be saved if his wood had been previously well seasoned under shelter.

The loss in using green fuel is, however, sometimes overrated. It has been found by experiment that one pound of the best seasoned wood is sufficient to heat 27 lbs. of water from the freezing to the boiling point.* This will be equal to heating and evaporating four pounds of water by every pound of wood. The 25 cubic feet of water, therefore, in every cord of green wood, weighing about 1,500 pounds, would require nearly 400 pounds of wood for its evaporation, or about one-seventh or one-eighth of a cord. Hence we may infer that seven cords of dry wood are about equal to eight cords of green. This imperfect estimate will apply only to the best hard wood, and will vary exceedingly with the different sorts of fuel; the more porous the wood becomes, the greater will be the necessity for thorough seasoning.

---

* The following results show the heating power of several combustibles:

1 lb. of wood (seasoned, but still holding 20 per cent of water) raised from 32° to 212°.............................	27 lbs. water.
1 lb. of alcohol...	68 " "
1 lb. of charcoal..	78 " "
1 lb. of oil or wax...	90 " "
1 lb. of hydrogen..	216 " "

It should be remembered that by ordinary modes of heating water, a very large proportion of the heat is wasted by passing up the chimney and into surrounding bodies, and the air.

Superficial observation often leads to very erroneous conclusions. Seasoned wood will sometimes burn with great rapidity, and, producing an intense heat for a short time, will favor an overestimate of its superiority. Green wood, on the other hand, kindles with difficulty, and burns slowly and for a long time; hence, where the draught of the chimney can not be controlled, it may be the most economical, because a less proportion of heat may be swept upward than by the more violent draught produced from dry materials. Where the draught can be perfectly regulated, however, seasoned wood should be always used, for convenience and comfort, and for economy.

Where wood is to be drawn to a distance, the preceding estimate shows that the conveyance of more than half a ton of water is avoided in every cord by seasoning.

CHAPTER II.

RADIATION OF HEAT.

The passage of heat through conducting bodies has been already explained. There is another way in which it is transmitted, termed *radiation*, in which it is thrown off instantaneously in straight lines from hot bodies, in the same way that light is thrown off from a candle. A familiar instance is furnished by the common or open *fire-place*, before which the face may be roasted with the radiated heat, while the back is chilled with cold. A screen held in the hand will intercept this radiated heat, showing that it flies in right lines like the rays of light.

Radiated heat is reflected by a polished metallic surface,

in the same way that light is reflected by a looking-glass. A plate of bright tin held near the fire will not for a long time become hot, the heat being reflected from it without entering and heating it. But if it be blackened with smoke, it will no longer reflect, but absorb the heat, and consequently will speedily become hot. This experiment may be easily tried by placing a new tin cup containing water over a charcoal fire, which yields no smoke. The heat will be reflected into the fire by the tin, and the water will scarcely become warm. But if a few pine shavings be thrown on this fire, to smoke the surface of the tin, it will then absorb the heat rapidly, and soon begin to boil. This explains the reason that bread bakes more slowly in a new tin dish, and that a polished andiron before a fire is long in becoming hot.

A concave burning-mirror, which throws the rays of heat to a focus or point, may be made of sheet-tin, by

Fig. 294.

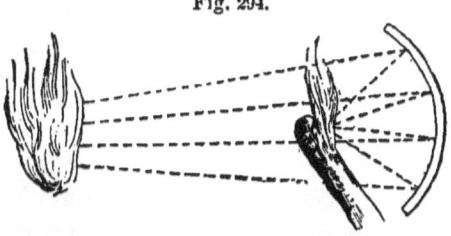

beating it out concave so as to fit a regularly curved gauge. If a foot in diameter, and carefully made, it will condense the rays of heat so powerfully at the focus, when held several feet from the fire, as to set fire to a pine stick or to flash gunpowder (fig. 294).

The reflection of radiated heat may be beautifully exhibited by using two such concave tin mirrors. Place them on a long table several feet apart, and ascertain the focus of each by means of the light of a candle. Then place in the focus of one a red-hot iron ball, or a small chafing-dish of burning charcoal. In the focus of the other place the wick of a candle with a small shaving of phosphorus in it. The heat will be reflected, as shown by

Fig. 295.

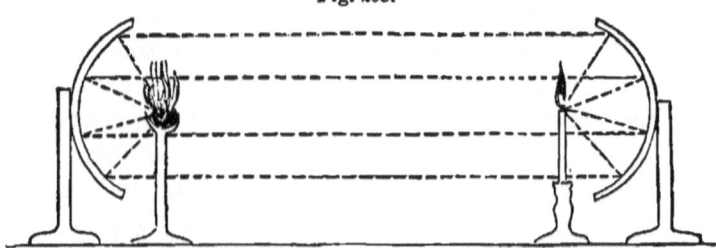

the dotted lines (fig. 295), and, setting fire to the phosphorus, will light the candle.

If a thermometer be placed in the focus of one mirror while the hot iron ball is in the other focus, it will rise rapidly; but if a lump of ice be substituted for the ball, the thermometer will immediately sink, and will continue to do so until several degrees lower than the surrounding air; because the thermometer radiates more heat to the mirrors, and then to the ice, than the ice returns.

DEW AND FROST.

All bodies are constantly radiating some heat, and if an equal amount is not returned by others, they grow colder, like the thermometer before the lump of ice. Hence the reason that on clear, frosty nights, objects at the surface of the earth become colder than the air that surrounds them. The heat is radiated into the clear space above without being returned; plants, stones, and the soil thus become cooled down below freezing, and, coming in contact with the moisture of the air, it condenses on them and forms *dew*, or freezes into *white frost*. Clouds return or prevent the passage of the heat that is radiated, which is the reason there are no night-frosts in cloudy weather. A very thin covering, by intercepting the radiated heat, will often prevent serious injury to tender plants. Even a sheet of thin muslin, stretched on pegs over garden vegetables, has afforded sufficient protection, when those around were destroyed.

FROST IN VALLEYS.

On hills, where the wind blows freely, it tends to restore to plants the heat lost by radiation, which is the reason that hills are not so liable to sharp frosts as still valleys. When the air is cooled it becomes heavier, and, rolling down the sides of valleys, forms a lake of cold air at the bottom; this adds to the liability of frosts in low places. The coldness is frequently still further increased by the dark and porous nature of the soil in low places radiating heat faster to the clear sky than the more compact upland soil.

A knowledge of these properties teaches us the importance of selecting elevated places for fruit-trees, and all crops liable to be cut off by frost; and it also explains the reason that the muck or peat of drained swamps is more subject to frosts than other land on the same level. Therefore, corn and other tender crops upon such porous soils must be of the earliest ripening kinds, so as to escape the frosts of spring by late planting, and those of autumn by early maturity.

REMARKABLE EFFECTS OF HEAT ON WATER.

The effects of heat and cold on water are of a very interesting character. Without its expansion in freezing, the soil would not be pulverized by the frost of winter, but would be found hard, compact, and difficult to cultivate in spring; without its expansion into steam, the cities which are now springing up, and the continents that are becoming peopled, through the influence of rail-ways, steam-ships, and steam manufactures, would mostly remain unbroken forests; without the crystallization of water, the beautiful protection of plants by a mantle of snow, in northern regions, would give place to frozen sterility; without the conversion of heat to a latent state in melting, the deepest snows would disappear in a moment from the earth, and cause disastrous floods; without its conversion to a latent state in steam, the largest vessel of boiling water would instantly flash into vapor. All these facts show that an extraordinary wisdom and forethought planned these laws at the creation; and even what appears at first glance as an almost accidental exception in the contraction of bodies by cold, and which causes ice to float upon water, preventing the entire masses of rivers and lakes from becoming permanently frozen, furnishes one out of an innumerable array of proofs of creative design in fitting the earth for the comfort and sustenance of its inhabitants.

APPENDIX.

SIMPLE APPARATUS FOR ILLUSTRATING MECHANICAL PRINCIPLES.

For the assistance of lecturers, teachers, and home students, the following list is given of cheap and simple apparatus and materials for performing most of the experiments described in the first part of this work. These experiments, although simple, exhibit principles of much practical importance.

1. Inertia apparatus, p. 12. The concave post or stand is sufficient, the snapping being done by the finger, although a spring-snap performs the experiment more perfectly.

2. Weight with two hooks and fine thread, p. 13.

3. The inertia of falling bodies may be simply shown, and the pile-engine illustrated, by placing a large wooden peg or rod upright in a box of sand, and then dropping a weight upon its head at different heights, which will drive the rod into the sand more or less, according to the distance passed through by the falling weight.

4. A straw-cutter, so made that the fly-wheel can be easily taken off, will show in a very striking manner the efficacy of this regulator of force.

5. Two lead musket balls will exhibit the experiment in cohesion, p. 27. Balls or lead weights with hooks may be separated by suspending weights, to show the amount of force required to draw them asunder. Metallic buttons or plates an inch in diameter, with hooks, will show the great strength needed to separate them when coated with grease, p. 27.

6. Capillary tubes of different sizes, two straight small panes of glass, and a vessel of water, highly colored with cochineal or other dye, to exhibit capillary attraction.

7. Glass tube, piece of bladder, and alcohol, for experiment described on p. 33.

8. The cylinder for rolling up the inclined plane, represented by fig. 18, p. 34, may be very easily made by using a round pasteboard box a few inches in diameter, and securing a piece of lead inside by loops made with a needle and thread. The object shown by fig. 19 may be cut in one piece out of a pine shingle, the centre rod being lengthwise with the grain; the two extremities are shaved small, and wound with thick sheet-lead, and the whole then colored or painted a

dark hue, to render the lead inconspicuous. The experiment with the penknives, p. 35, is very simple, care being taken to insert them low enough in the stick.

9. Irregular pieces of board, variously perforated with holes, and furnished with loops to hang on a pin, may be used to determine the centre of gravity, according to the principle explained by fig. 21, p. 35.

10. Portions of plank and blocks of wood, with the centre of gravity determined as in the last experiment, may have a plumb-line (which may be a thread and small perforated coin) attached to this centre, and then be placed on differently inclined surfaces, to show their upsetting just as this line of direction falls without the base. Toy-wagons, bought at the toyshops, may be variously loaded and used in experiments of this sort.

11. Experiments with the lever of the first kind may be easily performed by the use of a flat wooden bar, two or three feet in length, marked into inches, and placed on a small three-cornered block as a fulcrum. Weights, such as are used for scales, may be variously placed upon the lever. Levers of the second and third kind, which are *lifted* instead of borne down, may have a cord attached to the point where the power is to be applied, running up over a pulley or wheel, with a weight suspended to the other end.

12. An axle, furnished with wooden wheels with grooved edges, of different sizes, may be used to exhibit the principle of the wheel and axle, in connection with scale-weights that are furnished with hooks. The power of combined cog-wheels may be shown by a combination like that represented on p. 57, using weights for both cords.

13. Interesting experiments with the inclined plane, at different degrees of slope, by a contrivance similar to that represented by fig. 96, p. 83, with the addition of a small wheel at the upper side for a cord to pass over. This cord is fastened at one end to a light toy-wagon, running up and down the plane, and at the other to a weight suspended perpendicularly just beyond the upper edge of the plane. The wagon is variously loaded with weights, to counterpoise the suspended weight at different degrees of inclination.

14. A lecturer may quickly demonstrate before a class the small increase in the length of a road, in consequence of a considerable curve to one side of a straight line (as shown by fig. 69), by using a cord for measuring, the diagram being marked on a board or the wall.

15. A round stick of wood, and a long, wedge-shaped slip of paper, easily show the principle of fig. 75, p. 70.

16. A cog-wheel with endless screw and winch (fig. 77, p. 71), exhibits distinctly the great power of the screw in this combination.

17. Pine sticks, two feet long, and one-fourth to one-half inch through, of different shapes and sizes, supported at each end, and with weights hung at the middle till they break, may be made to illustrate the principles described on pp. 80, 81.

18. Some of the principles of draught may be shown, and especially

those in relation to the different angles of inclination for hard and soft roads, by using a common spring-balance as a dynamometer, attached to a hand-wagon, and also to a sliding block of wood.

19. Bent glass tubes, with arms of different sizes, to indicate the upward pressure of liquids, may be procured cheaply at glass-works. The experiment described by fig. 231, p. 204, may be rendered easy and interesting by purchasing a large and perfectly-working *syringe*, and attaching to its nose, by means of sealing wax, a slender glass tube two or three feet long. Fill the syringe with water, leaving the tube empty; then, with the tube upright, drive the water up through it with the piston of the syringe, and the increased weight felt on the piston as the column of water rises will be very evident.

20. A hydrostatic bellows a foot in diameter, made by any good mechanic, will answer the purpose well, and exhibit an important principle.

21. Specific gravities may be shown before a class by a common balance and a fine cotton or silk thread.

22. A tin pail, with a hole half an inch or an inch in diameter at the bottom, will show the contracted stream which pours from it, p. 212. A short tin tube, with a slight flange at the upper end (quickly made by any tin-worker), fitted into this hole, will increase the discharge, as shown by figs. 236, 237, and the difference in time for emptying the vessel may be measured by a stop-watch.

23. Archimedes' screw is readily made by winding a lead pipe round a wooden cylinder.

24. A glass syphon, filled with cochineal water, shows distinctly the theory of waves, by blowing with the mouth into one end.

25. Any vessel, filled with sand which has been heated over a fire, with rods of different substances, nearly of an equal size and length, and thrust with one end into the hot sand, in an inclined or nearly horizontal position, will exhibit the various conducting powers of these rods by melting pieces of wax or tallow placed on the ends most remote from the sand.

26. The expansion by heat may be demonstrated by fitting an iron rod to a hole in sheet-iron; on heating the bar it can not be made to enter. Or, if a hot iron ring be slipped on a tapering cold iron rod, it will contract on cooling so that the force of a man can not withdraw the rod.

27. The rising and descending currents in a vessel of heating water are easily rendered visible by throwing into a glass vessel, or flask, over a lamp, particles of sawdust from any hard, green wood, whose specific gravity is about the same as that of water.

28. Instrument figured on p. 265, for showing the principle of the steam-engine.

29. Experiments in latent heat may be easily exhibited with the assistance of a common thermometer.

30. Tin mirrors for showing radiation, p. 278.

DISCHARGE OF WATER THROUGH PIPES.

Table showing the amount of water discharged per minute through an orifice one inch in diameter; also through a tube one inch in diameter and two inches long, according to experiment. To ascertain the amount in gallons, divide the cubic inches by 231.

Height of head of water.	Amount discharged through Orifice.	Amount discharged through Tube.
1 Paris foot*	2,722 cub. in.	3,539 cub. in.
2 "	3,846 "	5,002 "
3 "	4,710 "	6, 6 "
4 "	5,436 "	7,070 "
5 "	6,075 "	7,900 "
6 "	6,654 "	8,654 "
7 "	7,183 "	9,340 "
8 "	7,672 "	9,975 "
9 "	8,135 "	10,579 "
10 "	8,574 "	11,151 "
11 "	8,990 "	11,693 "
12 "	9,384 "	12,0 5 "
13 "	9,764 "	12,699 "
14 "	10,130 "	13,177 "
15 "	10,472 "	13,620 "

VELOCITY OF WATER IN PIPES.

The following table shows the height of a head of water required to overcome the friction in horizontal pipes 100 feet long, and to produce a certain velocity, according to SMEATON:

Bore of Pipes.	6 Inches.	1 foot.	1½ feet.	2 feet.		3 feet.		4 feet.		5 feet.	
in.	in.	in.	in.	ft.	in.	ft.	in.	ft.	in.	ft.	in.
½	4.5	16.7	35.1	4	9.7	10	1.0	17	10.0	28	0.2
¾	3.0	11.1	23.3	3	2.5	6	8.6	11	10.6	18	8.1
1	2.2	8.4	17.5	2	4.9	5	0.5	8	11.0	14	0.0
1¼	1.8	6.7	14.0	1	11.1	4	0.4	7	1.6	11	2.5
1½	1.5	5.6	11.7	1	7.2	3	4.3	5	11.3	9	4.1
1¾	1.3	4.8	10.0	1	4.5	2	10.6	5	1.1	8	0.1
2	1.1	4.2	8.7	1	2.4	2	6.2	4	5.5	7	0.0
2¼	1.0	3.7	7.8	1	0.8	2	9.9	3	11.6	6	2.7
2½	0.9	3.3	7.0	0	11.5	2	0.2	3	6.8	5	7.2
3	0.7	2.8	5.0	0	9.6	1	8.2	2	11.7	4	8.0
3½	0.6	2.4	5.0	0	8.2	1	5.3	2	6.6	4	0.0
4	0.6	2.1	4.4	0	7.2	1	3.1	2	2.7	3	6.0

---

* A Paris foot is about 12 4-5 U. S. inches, and 15 Paris feet are about 16 U. S. feet.

RULE FOR THE DISCHARGE OF WATER.

Look for the velocity of the water per second in the pipe, in the upper line; and in the column beneath it, and opposite the given diameter of the pipe, is the height of the column or head required to obtain the required velocity.

To find the quantity of water discharged each minute, multiply the velocity by 12, which will give the inches per second; then multiply this product by 60, which will give the inches per minute; then, to change these cylindrical inches into cubic inches, multiply by 4 and divide by 5.* Divide the cubic inches by 231, and the result will be gallons.

By comparing this table with the next preceding, we shall perceive that the water flows from three to four times as fast through the tube two inches long, as through a tube one hundred feet long, the diameter of the tube and the head of water being the same.

RULE FOR THE DISCHARGE OF WATER.

The following general formula or rule, applicable to different cases, has been furnished by a practical engineer. It may be useful in ascertaining the quantity required to fill the driving pipe of a water-ram, and for various other purposes occasionally occurring in practice.

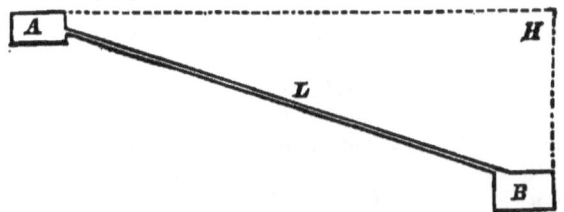

Let A represent the fountain or reservoir from which water is to be conveyed to the trough B through the pipe L. Let H be the height of the surface of the water in the reservoir, above the place of discharge, L the length of the tube in feet, and let D be the diameter of the tube in the smallest part. It is required to find the quantity, Q, which will be discharged in a second of time. The length and height being given in feet, and the diameter of the tube in inches, the formula, when the quantity is required in gallons, is as follows:

$$Q = 0.608 \sqrt{(D^5 \frac{H}{L})}$$

---

* This gives the cubic inches very nearly; but, to be more accurate, multiply the decimal .7854, which represents the difference between the area of a square and of a circle.

In order to make the above formula more intelligible:

Let L = 80 rods or 1320 feet.
" H = 50 feet.
" D = 2 inches.
" Q = gallons.

Then $Q = 0.608 \sqrt{(32 \times \frac{50}{1320})} = 0.67$; or, the same may be thus expressed in words:

Divide the height (50) by the length (1320); multiply the quotient by the fifth power of the diameter (fifth power of $2 = 32$); extract the square root of the product, which, being multiplied by 0.608, will give (0.67) the number of gallons the tube will discharge in one second; which, in this case, is 40 gallons in one minute.

VELOCITY OF WATER IN TILE DRAINS.

An acre of land in a wet time contains about one thousand spare hogsheads of water. An underdrain will carry off from a strip of land about two rods wide, and one eighty rods long will drain an acre. The following table will show the size of the tile required to drain an acre in two days' time, (the longest admissible), at different rates of descent, or the size for any larger area:

Diameter of Bore.	Rate of Descent.	Velocity of Current per second.	Hogsheads discharged in 24 hours.
2 inches.	1 foot in 100	22 inches.	400
2 inches.	1 foot in 50	32 inches.	560
2 inches.	1 foot in 20	51 inches.	900
2 inches.	1 foot in 10	73 inches.	1290
3 inches.	1 foot in 100	27 inches.	1170
3 inches.	1 foot in 50	38 inches.	1640
3 inches.	1 foot in 20	67 inches.	3100
3 inches.	1 foot in 10	84 inches.	3600
4 inches.	1 foot in 100	32 inches.	2500
4 inches.	1 foot in 50	45 inches.	3500
4 inches.	1 foot in 20	72 inches.	5600
4 inches.	1 foot in 10	100 inches.	7800

A deduction of one-third to one-half must be made for the roughness of the tile or imperfection in laying. The drains must be of some length, to give the water velocity, and these numbers do not, therefore, apply to very short drains.

GLOSSARY

OF TERMS USED IN MECHANICS AND FARM MACHINERY.

AXIS, a real or imaginary line, passing through a body, on which it is supposed to revolve.

AXLE or AXLE-TREE, the bar of metal or timber, on the ends of which the wheels of a carriage or wagon or other wheels revolve.

BABBETT METAL, an alloy, usually of tin and copper, for casing the supports of journals, either for repair, or for easier running.

BACK FURROW, to throw the earth from two plow-furrows together.

BALL-COCK, a self-regulating stop-cock, closed or opened by the rising or falling of a floating hollow ball.

BALL-VALVE, a valve consisting of a loose ball, fitting closely, prevented from moving beyond a certain limit.

BAND-WHEEL, a wheel in machinery on which a band or belt runs.

BEAM, the main lever of a steam-engine, turning on the centre, with the piston rod at one end, and the working-rod at the other. Also, the main timber or bar of a plow.

BEARING, the part of a shaft or spindle which is in contact with the supports.

BED, the foundation on which a fixed machine rests, as "the *bed* of an engine."

BELL-CRANK, a crank resembling that by which the direction of a bell-wire is changed.

BEVEL-GEAR, the gearing of cog-wheels placed obliquely together, or with the two axes forming an angle.

BOLSTER, the cross-bar of a wagon, resting on the axle, holding the box, and through which the king-bolt passes.

BRAKE, a lever or other contrivance used for retarding the motion of a wheel by friction against it.

BREAST-WHEEL, a water wheel where the current is delivered upon it about one-half or two-thirds its height, which distinguishes it from undershot and overshot wheels.

BRIDLE, the forward iron on the beam of a plow, to which the team is attached.

BRUSH-WEEEL, a wheel in light machinery, turned by friction merely, instead of cogs; bristles or brushes being often fixed to them to increase the friction of their pressing surfaces.

BUSH, the hollow box fitted into the centre of a wheel to take the bearing of an axle or journal.

CAM, the projecting part of an eccentric or wavy wheel, to produce alternate or reciprocating motion.

CANT-HOOK, a wooden lever with an iron hook near one end, used for moving heavy articles, particularly saw-logs, etc. The end of the lever is usually placed on the fulcrum, and the hook is fixed into the weight, making it a lever of the second kind.

CAPILLARY ATTRACTION, the attraction which causes liquids to rise in very small tubes, or which retains water among sand and the particles of soil.

CENTRE OF GRAVITY, that point in a body or mass of matter, around which all parts exactly balance each other.

CENTRIFUGAL FORCE, tending to fly from the centre, as the stone from a sling.

CENTRIPETAL FORCE, drawing towards the centre, like the cord of a sling.

CHAMFER, a slope, channel, or groove, cut in wood or metal.

CHASE, a wide groove.

CHILLED, applied to cast-iron rendered harder by casting the melted metal against cold metal in the mould, for rendering certain parts harder which are most liable to become worn.

CHINE, the ends of the staves of a barrel, outside the heads.

CLAMP, a cross-bar used to give additional strength, or to prevent warping. Also, a piece of metal or wood, generally resembling in shape the letter U, furnished with a screw, to fasten objects to a table or other fixed bodies, or to each other.

CLEAT, a piece of wood nailed across, to give strength or security.

CLEVIS, a draught iron, usually somewhat in the form of a bow or letter U, placed on the forward end of a plow-beam for draught, or for similar purposes.

CLICK, a pawl, a latch, or the ratchet of a wheel.

COG, the tooth or projection of a cog-wheel.

COLLAR, a metal ring around the end of a cylinder of wood to prevent splitting, or a ring around a piston or a journal, for securing tightness or steadiness.

COLTER or COULTER, the upright cutting iron of a plow.

COMPASS, an instrument for describing circles, measuring distances, etc.

COUNTER-SINK, a cavity made to receive the head of a screw.

COUPLING-BOX, a contrivance for connecting shafts, or throwing wheels in and out of gear.

CRAB, a small portable crane.

CRADLE, a scythe with fingers, for cutting grain by hand.

CRANE, a machine for raising weights and then swinging them sidewise; generally made by attaching a pulley to a swinging bar or frame.

CRANK, an axle with a crooked portion for changing a rotary to an alternate motion, or the reverse. A *three-throw* crank has three bends, for driving three pumps, each stroke separated from the others by the third of a revolution, thus making a regular and uniform application of the force.

CROSS-CUT SAW, a large saw worked by a man at each end, for cutting logs.

CUTTER-BAR, the cutting apparatus of a mowing or reaping machine.

CYCLOID, a curve made by any point in a circle rolling on a straight line, and marking the curve on a plane surface at the side of the circle. A rail driven in the rim of a wagon-wheel, driven through a snow bank, will mark a cycloid on the snow. An *epicycloid* is made by a similar revolution of a circle, rolling on the circumference of another circle, externally or internally.

DEAD CENTRE, a centre which does not revolve.

DEAD FURROW, the furrow where the plow throws the earth in opposite directions, or where the furrows meet in plowing a strip of land.

DERRICK, a pole or upright timber for supporting a crane, used in lifting heavy materials in building and for other purposes.

DOG, an iron catch or clutch, driven into the end of a saw-log, to hold it in a fixed position while sawing.

DOUBLE-TREE, the central whiffle-tree of a two-horse set.

DOWEL, a short iron or wooden pin to join two pieces of timber, protecting from one timber into a hole in the corresponding one. A familiar example occurs in the manner in which a cooper secures two or more boards in forming the head of a cask.

DRAUGHT, ANGLE OF, the angle made by a line of draught with a line drawn on the surface over which the body is drawn.

DREDGE, or DREDGING MACHINE, a machine for scooping up mud or earth from under water, for clearing the channels of canals, rivers, and harbors.

DRILL, a furrow for the reception of seed, or a row of growing plants; also a machine for sowing seed in continuous rows.

DRIVING-WHEEL, the wheel of a mowing or reaping machine, which runs on the ground, and propels the gearing.

DRUM, a revolving cylinder, around which belts or endless straps are passed, to communicate motion.

DYNAMICS, the science of motion and forces.

DYNAMOMETER, an instrument for measuring forces, applied to plows, mowing machines, thrashing machines, etc., to show the amount of force required to work them.

ECCENTRIC, out of centre; applied to wheels, discs, or circles, with the axle out of centre, to create reciprocating motion. *Eccentric rod* is the rod that transmits the motion of the eccentric wheel.

ELEVATOR, an endless revolving leather strap, set with sheet-iron boxes or cups, for raising grain. The term is also applied to buildings into which grain is thus elevated and stored.

EMERY WHEEL, a wheel set with emery at the circumference, for grinding or polishing metals.

ENDLESS CHAIN, a chain with the ends connected together, running on two drums or cylinders; as in the endless chain or tread-powers to thrashing machines.

ENDLESS SCREW, a screw working in a toothed wheel or cog-wheel, and imparting a motion to the wheel equal to the advance of one tooth to each revolution of the screw.

EPICYCLOID, see CYCLOID.

EPICYCLOIDAL WHEEL, a wheel with cogs on its interior rim, fitting into another cog-wheel precisely one-half its diameter, for converting circular into alternate motion; any point in the circumference of the smaller wheel, while in motion, describing a straight line.

EVENER, the central or larger whiffle-tree of a set of whiffle-trees for two horses, called also a *double-tree*.

FAN, the vane of a wind-mill, to keep the sails facing the wind.

FEATHER, the thin cutting part of a plowshare, on the right-hand side.

FELLOE, or FELLY, the circumference or rim of a wheel, or a segment of it, into which the spokes are inserted.

FERRULE, a ring or band on the end of a wooden rod or bar, to prevent splitting.

FEMALE SCREW, a hole cut with the threads of a screw, into which a screw fits.

FINGER-BAR, that portion of the cutting-bar of a mowing or reaping machine, in which the knife-bar works.

FLANGE, a projection from the end of a pipe or from any piece of mechanism, so as to screw to another part; a term also applied to the projection of a car-wheel to keep it from running off the rail.

FLASH-WHEEL, a water-wheel used for elevating water, resembling a breast-wheel with a reversed motion.

FLOAT-BOARD, one of the boards forming the exterior of a water-wheel, against which the stream of water dashes.

FLUME, the water passage of a mill, usually a box of plank.

FLY-WHEEL, a wheel with a heavy rim, for retaining inertia and equalizing the motion of machinery.

FOOT-VALVE, a valve in a steam-engine, opening from the condenser towards the air-pump.

FORCE-PUMP, or FORCING-PUMP, a pump with a solid piston, which drives instead of sucking water.

FRICTION-WHEEL, made by two wheels overlapping each other, and bearing between them the axle or journal of another wheel, thus diminishing the friction of the latter.

FULCRUM, a support; applied to the support used for the lever, in raising weights.

FURROW SLICE, the strip of earth thrown out by the plow at a passing.

FURROWS, FLAT and LAPPING; when the slice is laid flat or level, and when the edge of one overlaps the preceding, respectively.

GANG-PLOW, a compound plow made of a series of plows running side by side.

GAVEL, a sheaf of grain reaped but not bound,

GEARING, a series of cog-wheels working together.

GOVERNOR, a self-regulator of a steam-engine, so constructed that centrifugal force throws up weights when the engine runs too fast, and partly closes the admission pipe of steam; and, dropping again when it runs too slow, opens the steam pipe.

GRAVITATION, the attraction between bodies *in mass*, as distinguished from *cohesion* between the *particles;* the force which causes bodies to fall by the attraction of the earth.

GUARD, one of the fingers in the cutting apparatus of a mowing or reaping machine, for protecting the knives from injury from external objects. *Open Guard* has an opening above the knives, to prevent clogging.

HAY-TEDDER, a machine for spreading and turning hay.

HEADER, a reaping machine which cuts the heads of the grain and leaves most of the straw standing.

HEAD-LAND, the strip or border of unplowed land left at the ends of the furrows.

HOUND, the forward portion of a wagon, to which the tongue is attached.

HYDRAULIC RAM, see RAM.

HYDRAULICS, the science of *water in motion*, or the laws of motion and force as applied to running water, and to machinery driven by it.

HYDRODYNAMICS, the laws of motion and force, as applied to liquids, both in motion and at rest, and embracing *Hydraulics* and *Hydrostatics*.

INCLINED PLANE, a plane or surface deviating more or less from a level.

INERTIA, the property or force of matter by which it retains its state of motion or rest,—requiring force to start a body at rest or to stop one in motion.

JACK, an engine or machine for raising heavy weights.

JACK-SCREW, a strong iron screw for raising timbers, buildings, etc.

JOURNAL, the portion of a shaft or axle which revolves on a support.

KERF, the opening or slit made by the passage of a saw.

KEY, a wedge of wood or metal driven into a mortise or opening, to secure two parts together.

KNEE-JOINT, or TOGGLE-JOINT, a contrivance for exerting power or pressure, by straightening a double bar with a joint like that of the knee.

LAND, a term applied to the oblong portion of a field around which the team passes in plowing, the field being usually divided into several *lands* for this purpose. The term is also applied to the side of a plow opposite the mould board, and a plow is said to *run to land* when it takes too wide a furrow-slice.

LAND-SIDE, the side of that portion of a plow which runs in the soil, opposite the mould-board, and next the unplowed portion of ground.

LANTERN-WHEEL, a pinion made of two small wheels connected by parallel rods which form the teeth.

LEVER, a bar or rod for raising weights, resting on a point called a *fulcrum*.

LEVER-POWER, see SWEEP-POWER.

MALE SCREW, a screw with a spiral thread, fitting into a hole with corresponding threads called a *female screw*.

MECHANICAL POWERS, the simple machines or elements of machinery, consisting essentially of the Lever and Inclined Plane; the lever comprising the Wheel and Axle and the Pulley, and the inclined plane comprising the Wedge and the Screw.

MASH, or MESH, to interlock, as the teeth of cog-wheels.

MECHANICS, the science that treats of forces and powers, and their action on bodies, and particularly as applied to the construction of machines.

MITRE, to cut to an angle of 45 degrees, so that two pieces joined shall make a right angle.

MOMENTUM, impetus; the force of a moving body.

MONKEY, an apparatus for disengaging and securing again the ram of a pile-engine.

MORTISE, a hole cut to receive the end or tenon of another piece.

NUT, a piece of iron furnished with a screw-hole, used on the end of a screw for securing the parts of machinery.

OVERSHOT WHEEL, a water-wheel, the circumference of which is furnished with cavities or buckets, into which the stream of water is delivered at the top, turning the wheel by its weight.

PALL, or PAWL, the catch of a ratchet-wheel; a click.

PENT-STOCK, an upright flume.

PERAMBULATOR, a measurer of distances, consisting of a wheel, and index to show by wheelwork the number of its turns.

PERCUSSION, CENTRE OF, that point of a moving body at which its impetus is supposed to be concentrated.

PILE-DRIVER, or PILE-ENGINE, an engine for driving piles into the ground, effected by repeatedly dropping a heavy weight on the heads of the piles; used mostly in swamp or water when the bottom is mud.

PINION, a small-toothed wheel, working in the teeth of a larger one.

PITCH, the distance between the centres of two contiguous cog-wheels.

PITCH LINE, the circle, parallel with the circumference, which passes through the centres of the teeth of a wheel.

PITMAN, a rod connected with a wheel or crank, to change rotary to reciprocating motion, or the reverse.

PLANET-WHEELS, two elliptical wheels connected by teeth running into each other, and revolving on their foci.

PLOW-BEAM, the main timber of a plow, by which it is drawn.

PLOW-SHARE, the front part beneath the soil, which performs the cutting—sometimes called *plow-shoe*, or *plow-point*.

PLUNGER, the piston of a forcing pump.

PNEUMATICS, the science treating of the mechanical properties of air.

POLE, the tongue of a reaping or other machine.

POWER, the moving force of a machine, as opposed to the weight, load, or resistance of the substance wrought upon; also called *prime mover*.

PROJECTILE, a body thrown through the air.

PULLEY, one of the mechanical powers, consisting of a grooved wheel called the *sheave*, over which a rope passes; the box in which the wheel is set is called the block. The term is also applied to a fixed wheel over which a band or rope passes.

PUMP, a hydraulic machine for raising water; or one for withdrawing air. The handle is called the *brake*.

QUANTITY OF MOTION, the velocity of a moving body multiplied by its mass.

RABBET, to pare down the edge of a board or timber.

RACK, a straight bar cut with teeth or cogs, working into a corresponding cog-wheel or pinion which drives or follows it.

RAG-WHEEL, a wheel with teeth or notches, on which an endless or revolving chain usually runs. Also applied to a ratchet wheel.

RAKE-HEAD, the cross-bar of a rake, which holds the teeth.

RAM, HYDRAULIC RAM, or WATER-RAM, a hydraulic machine or engine for raising water to a height several times greater than that of the head of water, by employing the momentum of the descending current in successive beats or strokes.

RATCHET-WHEEL, a wheel cut with teeth like those of a saw, against which a click or ratchet presses, admitting free motion to the wheel in one direction, but insuring it against reverse motion.

REACH, the bar which connects the forward and rear axles of a wagon or carriage.

REAM, to bevel out a hole.

RECIPROCATING MOTION, alternate motion, or a movement backwards and forwards in the same path.

REEL, the revolving frame of a reaping machine, to throw the standing grain towards the knives.

RESOLUTION OF FORCES, dividing a force into two or more forces acting in different directions; rendering a compound force into its several simple forces.

RESULTANT, a force produced by the combination of two or more forces.

SAFETY VALVE, a valve opening outwards from a steam boiler, and kept down by a weight, permitting the escape of steam when the pressure reaches a certain point, regulated by the degree of weight. The term also applies to a valve opening inwards, and similarly regulated, to prevent the pressure of the atmosphere from crushing in the boiler when the steam cools and leaves a vacuum.

SCOOP-WHEEL, a water-wheel with scoops or buckets around it, against which the current dashes.

SCREW-BOLT, a bolt secured by a screw, or with a screw cut upon it.

SCREW-PROPELLER, an instrument for driving a vessel, by means of blades twisted like a screw, revolving beneath the water, the axis being parallel with the keel.

SECTION, one of the knives or blades on the cutter-bar of a mowing machine.

SELF-RAKER, a contrivance attached to a reaping machine, to throw off the cut grain in gavels, to obviate raking off by hand.

SHEARS, or SHEERS, two poles lashed together like the letter X, for placing under heavy poles, etc., in raising them; also to single vertical poles supporting pulleys, for a similar purpose.

SHEAVE, the wheel of a pulley set in a block.

SHOOT, or SHUTE, a passage-way down which grain, hay, or straw, is slid or thrown.

SIDE-DRAUGHT, the side pressure of a machine on the team which draws it, as distinguished from *centre draught*.

SINGLE-TREE, a single whiffle-tree, the cross-bar to which the traces of a horse are attached, as distinguished from a *double-tree*, or two-horse whiffle-tree.

SIPHON, or SYPHON, a bent tube for drawing off liquids; the column of liquid in the outer or longer leg overbalancing the inner column, and producing a current.

SKEIN, the iron casing of a wagon-axle on which the wheel runs.

SKIM-COULTER, a coulter of a plow so constructed as to pare the surface before the mould-board.

SKIM-PLOW, the small forward mould-board of a double Michigan or Sod-and-subsoil plow.

SLIDE-REST, the rest or support of the chisel in a turning lathe, made to slide along the frame for cutting successively the different parts of the work.

SLOT, a slit or oblong aperture in any part of a machine, to admit another part.

SNATH, the handle or bar to which the blade of a scythe is attached.

SOD, the slice of earth cut by the passing of a plow.

SOLE, the bottom plate under a horse-shoe tile, in draining.

SPINDLE, a small axle in machinery, as distinguished from a shaft or large axle.

SPIRIT-LEVEL, a glass tube containing alcohol with an air-bubble, hermetically sealed at both ends, the position of the bubble at the middle showing the tube to be level.

SPUR-WHEEL, or PINION, a cog-wheel with teeth parallel to the axle.

STANDARD, an upright supporting timber; the front upright bar in a plow to which the mould-board is fastened.

STEAM CHEST, a box attached to the cylinder of a steam-engine, in which the sliding valves work.

STIRRUP, an iron band encasing a wooden bar, for attaching to some other part.

STUD, a short, stout support.

SUBSOIL-PLOW, a plow running below the furrow of a common plow, for breaking up or loosening the subsoil or lower soil of a field.

SWAGE, to give shape to a substance by stamping with a die.

SWEEP-POWER, a horse-power for driving thrashing and other machines, where the horses are attached to a pole and walk in a circle.

SWINGLE-TREE, also called SWING-TREE, SINGLE-TREE, WHIPPLE-TREE, and WHIFFLE-TREE; the cross-bar to which traces are attached.

SWING-PLOW, a plow with no wheel under the beam.

SWIVEL, a ring and axis in a chain, to admit of its turning.

SWIVEL BRIDGE, a bridge which turns round sideways on its centre.

SWIVEL PLOW, a side-hill plow, or a plow with a reversible mould-board.

TACKLE, a pulley, or machine with ropes and blocks for raising heavy weights.

TAIL-RACE, the channel which carries off the water below a water wheel.

TEDDER, a machine for turning and spreading hay.

THILL, one of the shafts of a wagon between which the horse is put —often corrupted to *Fill*.

THROTTLE-VALVE, a valve which turns at its centre on an axis—generally used to regulate the supply of steam to the cylinder of a steam-engine.

THUMB-SCREW, a screw with its head flattened in the direction of its length, so as to be turned with the thumb and finger.

TIDE-WHEEL, a wheel adapted to currents flowing both ways—the float-boards pointing from the centre.

TINE, the tooth or prong of a fork.

TIRE, the iron band which binds together the fellies of a wheel.

TOGGLE-JOINT, or knee-joint, a mechanical power exerted by straightening a double bar with a hinge at the middle or connection.

TORSION, the act of twisting by the application of lateral force. The *force of torsion* is the elasticity of a twisted body.

TRACK-CLEANER, an attachment to a mowing machine, to throw the cut grass away from that which is uncut.

TRACTION, ANGLE OF, the angle between the line of draught and any given plane, as that of the earth's surface.

TRAMMEL, an instrument used by carpenters for drawing an ellipse.

TREAD-POWER, a machine on which the horse or other animal working it walks. It may be either a horizontal or slightly inclined wheel; or an endless-chain power, the term being more frequently applied to the latter.

TRENCH-PLOW, a plow cutting deep furrows and bringing the subsoil up to the surface; as distinguished from a subsoil plow, which only loosens the subsoil and leaves it below the surface.

TRUNDLE-HEAD, a wheel turning a mill-stone.

TUB-WHEEL, a horizontal water-wheel, driven by the percussion of the stream against its floats, and not submerged in water.

TUMBLER, a latch in a lock, which, by means of a spring, detains the bolt in its place until lifted by the key.

TUMBLING ROD, the rod which connects the motion of a horse-power with that of a thrashing or other machine.

TURBINE WHEEL, a horizontal water-wheel, so constructed that the current strikes all the floats or buckets around the circumference at the same time, thus imparting to it great power for its size. It is submerged, the water escaping towards the centre and below, or above and below together.

UNDERSHOT WHEEL, a water-wheel moved by the current striking against the lower portion of its circumference.

UNIVERSAL JOINT, a connecting joint between two rods, consisting of a sort of double hinge, admitting motion in any direction.

VALVE, a lid for closing an aperture or passage, so as to open only in one direction.

VELOCITY, speed or swiftness; which may be *uniform*, or equal throughout; *accelerated*, or increasing; or *retarded*, or rendered slower.

VIRTUAL VELOCITIES, PRINCIPLE OF, that by which certain powers are equal to each other, where the force and space moved over, whatever these may be, are the same when multiplied together.

WASHER, a circular piece of metal, pasteboard, or leather, placed below a screw-head, or nut, or within a linch-pin, for protection.

WATER-RAM, see RAM.

WHIFFLE-TREE, or WHIPPLE-TREE, the cross-bar to which the traces of a horse are attached; see SINGLE-TREE.

WHIP-SAW, a large saw, worked by a man at one end, with a wooden spring at the other; a cross-cut saw.

WINCH, a bent handle or right-angled lever, for turning a wheel or grindstone, or producing rotary motion for other purposes.

WINDLASS, a machine for raising heavy weights, by the winding of a rope or chain on a horizontal axle, and turned by a winch or by levers.

WINROW, or WINDROW, the ridge of hay raked up on a meadow.

WREST, a partition which determines the form of the bucket in an overshot wheel.

INDEX.

A

Air, Pressure of..................239
" Mode of weighing.............239
" Pump........................240
" Hand fastened by............241
" Motion of....................245
" Resistance of.................247
Alden's Cultivator...............146
Allen's Farm Mill................195
Altitudes measured by the Barometer....................243
American Hay-tedding Machine....165
Apparatus for Experiments.......281
Aqueducts of the Romans........199
Archimedean Root Washer........193
" Screw................217
Archimedes, would move the earth with a lever.................55
Artesian Springs and wells........201
Atmosphere, Height and Weight of.......................239, 241

B

Bags, How to carry................41
Balance, a lever...................47
Balls, Why they roll easily........38
Barometer........................241
Bars of wood, Strength of.........79
Beardsley's Hay Elevator..........177
Bellows, Hydrostatic..............204
Bevel Wheels or Bevel Gear........60
Billings' Corn Planter.............155
Binders for Reaping Machines.....163
Boat, Compound motion of........20
Broadcast Sower, Seymour's......154
Brown's Wind-mill................251
Brush Harrow....................142
Buckeye Mower...................159
Bullard's Hay-tedding Machine....165
Bulk of a ton of different substances.210
Burrall's Corn-sheller.............191

C

Capillary attraction...............31
" " its great importance...............32
Cayuga Chief Mower..............160
" " Dropper.............162
Cements, Effects of................28
Centre of Gravity..................31
" " curious examples of................!........35
" how determined......35
Centrifugal Force..................21
Chain Pump......................221
Cheese Press......................72
" " Dick's.................74
" " Kendall's.............73
Chimney Currents................253
" Caps..................254
Chimneys, Construction of........254
" To prevent smoking....256
Churn with fly-wheel..............17
" worked by dog-power......191
Cistern Pumps....................219
Cisterns, To calculate contents of..237
" Proper sizes for..........238
Clod Crusher.....................149
" " Croskill's and American..................150, 151
Cog, Hunting.....................60
Cogs, Form of....................58
" and Cog-wheels..............58
Cohesion, Attraction of............27
" between lead balls......27
" weak in liquids.........31
Complex Machines, objectionable..116
Compound motion................19
" " How to calculate. 20
Comstock's Rotary Spader....118, 117
Conducting power of bodies.......260
" " liquids......261
Corn Planter, Billings'.............155
" Sheller, Burrall's...........191

297   13*

Corn Sheller, Horse-power 192
" " Richards'............192
Corn Planters......................155
Cost of Implements and Machines.117
Cotton Gin, Emery's.........' 196
Coulter for Plows 127
Crested Furrow-slice...............126
Crosskill's Clod-crusher............150
Crow-bar, a simple power..........43
Crown Wheels......................60
Cubic foot of different substances,
    Weight of....................210
Cultivator, or Horse-hoe 145
    " Claw-toothed..........146
    " Alden's Thill.........146
    " Duck-foot 146
    " Two-horse.............148
    " Harrington's.........157
Cutter for the Plow................127
    " Bar in Mowers and Reapers.158

D

Dederick's Hay-press...............185
    " Capstan................185
Deep-tiller Plow, Holbrook's.......126
Deep Wells, Pump for. 220
Dew and Frost.....................278
Discharge of water through pipes..284
    " " Rule for........285
Ditches, Velocity of water in..214, 286
    " Leveling instruments for..115
Dog-power Churn...................101
Draught, Combined................. 96
Draught of wheels, explained...... 37
    " Line of...................93
    " Principles of.............93
    " How to measure........... 94
    " of Plows..................95
Drilling wheat....................153
Drills, Hand......................157
Drive-pump........................220
Dropper, attachment to reapers....162
Dynamometer, applied to roads.... 85
    " Construction and use
        of................ 98
    " Self-recording.......101
    " Waterman's.........102
    " for rotary motion...106

E

Elevators for Hay.................173
Emerson's Chimney Cap............235

Emery's Horse-powers..... 188
    " Cotton Gin...............196
Empire Wind-mill..................251
Endless-chain power........ .. 188, 189
Engine, Garden..:.................230
Experiments, apparatus for........281

F

Falling Bodies, Velocity of........ 23
    " " Resistance of air on 25
    " " in vacuo........... 25
Farm, Seventy-thousand-acre...... 8
    " implements, Construction
        and use of....................115
    " implements, Cost of.........117
    " mills........................195
Finger-bar in mowers and reapers..158
Flail, Old sort....................187
    " Estimate of comparative work
        with.........................187
Flash-wheel.......................231
Flea, power of leaping............115
Fly-wheel......................... 16
    " used on horse-pump..... 16
Forcing-pump......................223
Fork Handles, Proper form of...... 76
Forsman's Farm Mill...............195
Friction 81
    " Nature of.................. 82
    " How to Measure........... 83
    " not influenced by velocity. 88
    " of axles................... 89
    " of wheels.... 90
    " Lubricating substances for. 91
    " Advantages of............. 92
Frost and Dew.....................278
    " in valleys...................279
Fuel, Green wood for..............275
Furrow-slice, Crested.............126
Furrows, Lapping and flat........ 127

G

Galileo's experiment on falling bodies............................ 26
Garden Engine....................230
Garrett's Horse-hoe...............147
Geddes' Harrow...................143
Gladding's Hay-fork...............175
Glossary of terms.................287
Gravitation....................... 23
Gravity, Centre of...... 34

INDEX.

Gravity, Specific, how measured...208
" " of different substances......................209
Green wood for fuel...............274

H

Hand-drills........................157
" rakes, sulky.................160
Harrington's Seed-sower...........157
" Cultivator............157
Harrow, Norwegian................144
" Morgan..................144
" Scotch, or square.........143
Harvester, Marsh's.................163
Hay-forks, Horse..................173
" carriers......................180
" loaders.......................186
" rake, Revolving..............168
" " Warner's...............169
" rakes........................166
" " Simple.................167
" stacking machine.............184
" tedder, Bullard's..............165
" " American..............166
" presses......................184
" sweep.......................171
Headers............................163
Heat, Properties of................260
" Expansion by...........263, 271
" Latent.......................273
" Radiation of.................276
Hicks' Hay-carrier.................180
High pressure steam-engines.......269
Hoe-handle, Proper form of........ 77
Holbrook's Plow...................125
" Swivel or side-hill Plow.133
Horse, day's work at different degrees of speed...............110
" hoe, Garrett's..............147
" power, Estimating...........109
" Hay-forks, Operation of....174
" fork, Gladding's............175
" " Palmer's.............176
" " Myers'................177
" " Beardsley's...........177
" " Raymond's............178
" " Harpoon..............179
" " Walker's..............179
" " Sprout's...............179
Hydraulic Ram....................226
" " Regulating........227
Hydrostatic Paradox...............203

Hydrostatic Bellows...............204
" Press..................203
Hydrostatics......................198

I

Implements required for the farm.. 7
..............................9, 117
" Construction and use of.................115
Improvements in Farm Machinery. 8
Inclined Plane..................... 63
Inertia............................ 11
" apparatus................... 12
" Effects of, on wagons.....13, 17

J

Joint, Universal................... 60

K

Kirby Mower and Reaper..........159
" Reaper, Hand-rake for.......160
" " Self-raking..........161
Knee-joint, or Toggle-joint......... 71
Knives in mowers and reapers, Form of..........................158
Kooloo Plow......................118

L

Labor, Application of..............108
" of men and horses...........110
Ladders, Self-supporting........... 40
Lapping and flat furrows......127, 128
Latent heat........................273
" " Advantages of..........275
Law of virtual velocities........... 43
Leveling Instruments..............215
Levers............................ 45
" of the second kind.......... 45
" " " first kind............. 46
" " " third kind............ 46
" Calculating power of.....49, 50
" Examples of................. 46
" Combination of............. 50
Line of direction.................. 36
Liquids, Velocity of, in falling.....211
" Discharge through pipes..212
Loads on sideling roads........... 37
Lubricating substances............ 90

M

Machinery in connection with water.198
Machines, Advantages of........... 42
    " Models of................113
    " Complex, objectionable..116
    " Construction and use of..115
    " Required for the Farm.7, ...9, 117
Marsh's Harvester...................163
Materials, Measuring strength of... 29
Mechanical powers................. 42
    " principles, Advantages of..................... 10
Mechanical principles, Application of.............................. 75
Models of machines................113
Moline Plow....................... 120
Momentum.......................... 14
    " Calculating quantity of.. 18
    " of railway trains........ 18
Moorish Plow.......................118
Morgan's Harrow....................144
Motion, Compound.................. 19
Mouldboard of the Plow, Form of..124
Mountains, Height of, measured by barometer......................243
Mowing Machine, Wood's...........158
    " " Kirby's..........159
    " " Buckeye.........159
    " " Cayuga Chief....160
Mowing Machines, Construction of.158
    " " How to select... 164
Myers' Horse-fork..................177

N

Norwegian Harrow..................144

O

Ogle, inventor of the Finger-bar....159
Ox-yokes 78

P

Packer's Stone Lifter............... 62
Palmer's Horse-fork................176
    " Hay-stacking Machine....183
Paradox, Hydrostatic...............203
Pile Engine or Driver.............. 15
Pinions, Operation of.............. 60
Pipes, To determine strength of....200
    " Discharge of water through, ...213, 284

Pitts' Straw-carrier and Thrasher...190
Plank roads, Amount of resistance on............................84, 86
Planting Machines..................152
Plaster Sower, Seymour's..........155
Platform Scales..................52, 53
Plow, Kooloo.......................118
    " Moorish.......................118
    " German.......................119
    " Modern improved............119
    " Moline Steel..................120
    " Woodruff & Allen's..........120
    " Double Michigan.............131
    " Mole...........................139
    " Ditching......................138
    " Side-hill or Swivel...........132
    " Subsoil..................133, 135
    " Trench........................134
    " Paring........................137
    " Gang..........................137
    " Defects in....................122
    " Character of a good one......121
    " Cutting edge of...............121
    " Resistance of different parts..122
    " Form of the mouldboard.....124
    " Appendages to................140
    " Wheel coulter and Weed-hook on............................140
Plowing, Operation of..............128
    " Fast and slow.................130
    " Requisites for success in..129
Potato Planter, True's..............156
    " Digger........................144
Power of a horse, Estimating......110
Press, Hydraulic...................205
Presses for hay. 184
Pressure of liquids, Determining...202
    " Upward, Measuring.......199
    " " in liquids........198
Pulley............................. 61
Pulverizers........................142
Pump, Cistern.....................219
    " Non-freezing.................219
    " Drive.........................220
    " for deep wells...............220
    " Chain........................221
    " Rotary.......................222
    " Suction and Forcing........223
Pumping water by wind............218
Pumps, Construction of............218
Pyramids, Firmness of............. 38
Pyrometer, how made.............. 267

INDEX.

R

Rake, Simple form of............167
" Revolving.................168
" " Warner's...........169
" Spring-tooth.............170
" " " Hollingsworth's.171
Ram, Hydraulic....................226
Raymond's Hay Elevator............178
Reaping Machines during the war.. 8
" " Self-rakers for..161
" " Headers.........163
" " How to select...164
Revolving Hay Rake................168
Roads, importance of good ones.... 68
" How to form the bed of..... 67
" Measuring the friction on... 84
" Amount of resistance on.... 86
" Good and bad................ 69
" Ascent in.............63, 66
" Cost of going up and down hill.......................... 65
Rocks, Machines for removing...... 62
Rockers, How to make............. 41
Roners............................152
Rolling Mill, Principle of.......... 74
Root Washer......................193
" Slicers.....................194
Rotary Spader, Comstock's....148, 117
" Pump.....................232

S

Sack-barrow, a lever.............. 48
Sap, Ascent of.................... 33
Scotch or Square Harrow...........143
Screw............................. 70
" Archimedean................217
" Estimating power of......... 71
Seed Sower.......................153
" " Harrington's...........157
Self-raking Reapers...............161
Seymour's Broadcast Sower........154
Shares' Harrow...................145
Side-hill or Swivel Plow..........132
Single-tree, Wier's............... 98
Sowing Machines..................152
Specific gravities, how determined.208
" " Table of..........200
Springs of water..................201
Stacks, Building by machinery.....182
Steam engine, Construction of.265, 267
" " for farm purposes....270

Steel Plows.....................120
Steelyard......................... 47
Stone-lifter...................... 62
Straw-cutters.................16, 75
" carrier, Pitts'.............190
Strength of materials............. 29
" " wood, iron, and ropes. 30
" " rods and bars.......79, 80
" " pipes, To determine...200
Stubble Plow, Holbrook's..........126
Stump-puller...................... 54
Subsoil plowing...................133
" Plows......................135
Swivel Plow......................132
Syphon...........................244
" used for draining.........243

T

Teeth of wheels................... 58
Thill-cultivator, Alden's.........146
Thrashing by machinery...........187
" machine, Comparative cheapness of...............188
Thrashing machine, Endless-chain power for....................188
Thrashing machine, Pitts'.........190
Toggle-joint power................ 71
Tread horse-powers...............188
" " " To determine work of...................188
Turbine Water-wheel..............223
" " " Reynolds'.....224
" " " Van de Water's.......................224

U

Universal joint................... 60
Upward pressure of liquids........198

V

Vacuum, Machine running in....... 11
Velocity affects friction but slightly. 88
" of falling water...........211
" of water in ditches....214, 286
" " through pipes....284
Ventilation......................257
" through walls and garrets........................258
Ventilator, Griffith's............258
" Emerson's............255
Virtual velocities, Law or rule of... 43

W

Wagon springs, Advantages of..... 17
" wheels, Proper width for..... 87
Warner's Revolving Rake..........169
Washing Machine................... 72
Water, Remarkable effects of heat on ....,........279
Water, Velocity of..............211, 213
" Discharge of, through pipes.212
" in ditches....................214
" wheels, Turbine..............223
" ram........ 226
" engines..................... 230
Waves, Nature of..... 232
" Velocity of..................234
" Breadth and height of......233
" To prevent inroads of..235, 236
Weather glass..................... 243
Wedge............................ 69
Weed hook on plows..............140

Weighing machine, or platform scales................... ..52, 53
Wheat drill................... 152
" " Bickford & Huffman's, Construction of..............153
Wheel and axle.. 55
" " " Modifications of.... 57
Wheelbarrow, Operation of........ 47
Wheel-cutter to plows..............140
Wheels, large ones run best........ 39
" for wagons Proper width for........... 87
Whiffle-trees for three horses....50, 97
Wind, Causes of..... 252
" Velocity of.... 246
" mill....................... 247
" " Pumping water by.......248
" " Brown's..................251
Wooden legs, why hard to walk on. 40
Wood's Plow....................119
Work of men and horses, Estimating.......................110

www.ingramcontent.com/pod-product-compliance
Lightning Source LLC
Chambersburg PA
CBHW031248250426
43672CB00029BA/1377